Smart Innovation, Systems and Technologies

Volume 97

Series editors

Robert James Howlett, Bournemouth University and KES International,
Shoreham-by-sea, UK
e-mail: rjhowlett@kesinternational.org

Lakhmi C. Jain, University of Technology Sydney, Broadway, Australia;
University of Canberra, Canberra, Australia; KES International, UK
e-mail: jainlakhmi@gmail.com; jainlc2002@yahoo.co.uk

The Smart Innovation, Systems and Technologies book series encompasses the topics of knowledge, intelligence, innovation and sustainability. The aim of the series is to make available a platform for the publication of books on all aspects of single and multi-disciplinary research on these themes in order to make the latest results available in a readily-accessible form. Volumes on interdisciplinary research combining two or more of these areas is particularly sought.

The series covers systems and paradigms that employ knowledge and intelligence in a broad sense. Its scope is systems having embedded knowledge and intelligence, which may be applied to the solution of world problems in industry, the environment and the community. It also focusses on the knowledge-transfer methodologies and innovation strategies employed to make this happen effectively. The combination of intelligent systems tools and a broad range of applications introduces a need for a synergy of disciplines from science, technology, business and the humanities. The series will include conference proceedings, edited collections, monographs, handbooks, reference books, and other relevant types of book in areas of science and technology where smart systems and technologies can offer innovative solutions.

High quality content is an essential feature for all book proposals accepted for the series. It is expected that editors of all accepted volumes will ensure that contributions are subjected to an appropriate level of reviewing process and adhere to KES quality principles.

More information about this series at http://www.springer.com/series/8767

Ireneusz Czarnowski · Robert J. Howlett
Lakhmi C. Jain · Ljubo Vlacic
Editors

Intelligent Decision Technologies 2018

Proceedings of the 10th KES International
Conference on Intelligent Decision
Technologies (KES-IDT 2018)

 Springer

Editors

Ireneusz Czarnowski
Gdynia Maritime University
Gdynia
Poland

Robert J. Howlett
Bournemouth University
Poole
UK

and

KES International
Shoreham-by-Sea
UK

Lakhmi C. Jain
Centre for Artificial Intelligence, Faculty of
    Engineering and Information Technology
University of Technology Sydney
Sydney, NSW
Australia

and

Faculty of Science, Technology
    and Mathematics
University of Canberra
Canberra, ACT
Australia

and

KES International
Shoreham-by-Sea
UK

Ljubo Vlacic
Griffith Sciences - Centres and Institutes
Griffith University
South Brisbane, QLD
Australia

ISSN 2190-3018   ISSN 2190-3026   (electronic)
Smart Innovation, Systems and Technologies
ISBN 978-3-030-06352-8   ISBN 978-3-319-92028-3   (eBook)
https://doi.org/10.1007/978-3-319-92028-3

Printed on acid-free paper

This Springer imprint is published by the registered company Springer International Publishing AG
part of Springer Nature
The registered company address is: Gewerbestrasse 11, 6330 Cham, Switzerland

Preface

This volume contains the proceedings of the 10th International KES Conference on Intelligent Decision Technologies (KES-IDT 2018) held in Gold Coast, Queensland, Australia, on June 20–22, 2018.

KES-IDT is an international annual conference organized by KES International. The KES-IDT conference is a sub-series of the KES Conference series.

The conference provided opportunities for the presentation of new research results and discussion about them, leading to knowledge transfer and generation of new ideas in the field of intelligent decision-making.

This edition, KES-IDT 2018, attracted a number of researchers and practitioners from all over the world. The KES-IDT 2018 Program Committee accepted 24 papers for oral presentation and publication in the volume of the KES-IDT 2018 proceedings. These papers have been submitted for the main track and special sessions devoted to specific topics, such as:

- Decision-Making Theory for Economics,
- Advances in Knowledge-based Statistical Data Analysis,
- On Knowledge-Based Digital Ecosystems and Technologies for Smart and Intelligent Decision Support Systems,
- Soft Computing Models in Industrial and Management Engineering,
- Computational Media Computing and its Applications,
- Intelligent Decision-Making Technologies,
- Digital Architecture and Decision Management.

Each paper has been reviewed by 2–3 members of the International Program Committee and International Reviewer Board.

We are very satisfied with the quality of the program and would like to thank the authors for choosing KES-IDT as the forum for the presentation of their work. Also, we gratefully acknowledge the hard work of the KES-IDT International Program Committee members and of the additional reviewers for taking the time to review the submitted papers and selecting the best among them for presentation at the conference and inclusion in its proceedings.

We hope and intend that KES-IDT 2018 significantly contributes to the fulfillment of the academic excellence and leads to even greater successes of KES-IDT events in the future.

June 2018   Ireneusz Czarnowski
                                                                                      Robert J. Howlett
                                                                                       Lakhmi C. Jain
                                                                                        Ljubo Vlacic

KES-IDT 2018 Conference Organization

Honorary Chairs

Ljubo Vlacic	Griffith University, South Brisbane, Queensland, Australia
Lakhmi C. Jain	University of Canberra, Australia and Bournemouth University, UK
Gloria Wren-Phillips	Loyola University, USA
Junzo Watada	Waseda University, Japan

General Chair

Ireneusz Czarnowski	Gdynia Maritime University, Poland

Executive Chair

Robert J. Howlett	KES International and Bournemouth University, UK

Program Chair

Alfonso Mateos Caballero	Universidad Politécnica de Madrid, Spain

Publicity Chair

Izabela Wierzbowska   Gdynia Maritime University, Poland

Special Sessions

Decision-Making Theory for Economics

Eizo Kinoshita   Meijo University, Japan
Takao Ohya   Kokushikan University, Japan

Advances in Knowledge-Based Statistical Data Analysis

Mika Sato-Ilic   University of Tsukuba, Japan
Lakhmi C. Jain   Bournemouth University, UK/University
                                   of Canberra, Australia

On Knowledge-Based Digital Ecosystems and Technologies for Smart and Intelligent Decision Support Systems

Shastri L. Nimmagadda   Curtin Business School, Curtin University, Perth,
                                   Australia
Seema Purohit   Kirti College, Mumbai, India
Neel Mani   ADAPT Centre for Digital Content Technology,
                                   Dublin City University, Dublin, Ireland
Torsten Reiners   Curtin Business School, Curtin University, Perth,
                                   Australia

Soft Computing Models in Industrial and Management Engineering

Shing Chiang Tan   Multimedia University, Malaysia
Junzo Watada   Universiti Teknologi Petronas, Malaysia
Chee Peng Lim   Deakin University, Australia

Computational Media Computing and Its Applications

Ippei Torii   Aichi Institute of Technology, Japan
Takahito Niwa   Aichi Institute of Technology, Japan
Naohiro Ishii   Aichi Institute of Technology, Japan

Intelligent Decision-Making Technologies

Jeffrey W. Tweedale   DST Group, Australia

Digital Architecture and Decision Management

Alfred Zimmermann   Reutlingen University, Germany
Rainer Schmidt   Munich University, Germany

International Program Committee

Jair M. Abe   Paulista University, University of Sao Paulo,
                             Brazil
Witold Abramowicz   Poznan University of Economics and Business,
                             Poland
Alireza Ahrary   Faculty of Computer and Information Sciences,
                             Sojo University, Japan
Piotr Artemjew   University of Warmia and Mazury, Poland
Ahmad Taher Azar   Faculty of Computers and Information, Benha
                             University, Egypt
Hyerim Bae   Pusan National University, Korea
Valentina Emilia Balas   Aurel Vlaicu University of Arad, Romania
Alina Barbulescu   Ovidius University of Constanta, Romania
Dariusz Barbucha   Gdynia Maritime University, Poland
Andreas Behrend   University of Bonn, Germany
Mokhtar Beldjehem   University of Ottawa, Ottawa, Ontario, Canada
Monica Bianchini   Dipartimento di Ingegneria dell'Informazione
                             e Scienze Matematiche, Università degli Studi
                             di Siena, Italy
Gloria Bordogna   CNR IREA, Italy
Oliver Bossert   McKinsey & Co. Inc, Germany
János Botzheim   Szechenyi Istvan University, Hungary
Lars Brehm   University of Munich, Germany
Alfonso Mateos Caballero   Universidad Politécnica de Madrid, Spain
Frantisek Capkovic   Slovak Academy of Sciences, Bratislava,
                             Slovakia
Shyi-Ming Chen   National Taiwan University of Science
                             and Technology, Taipei, Taiwan
Adrian-Gabriel Chifu   Aix-Marseille Université (FEG/LSIS), France
Mario G. C. A. Cimino   University of Pisa, Italy
Marco Cococcioni   University of Pisa, Italy
Angela Consoli   Defence, Science and technology Group,
                             Australia

Paolo Crippa   Department of Information Engineering,
                                       Università Politecnica delle Marche, Ancona,
                                       Italy
Matteo Cristani   Universita di Verona, Italy
Alfredo Cuzzocrea   University of Trieste, Italy
Bogdan Cyganek   AGH University of Science and Technology,
                                       Poland
Ireneusz Czarnowski   Gdynia Maritime University, Poland
Eman El-Sheikh   University of West Florida, USA
Margarita N. Favorskaya   Reshetnev Siberian State University of Science
                                       and Technology, Russia
Michael Fellmann   University of Rostock, Germany
Bogdan Franczyk   Universitat Leipzig, Germany
Ulrik Franke   RISE SICS Swedish Institute of Computer
                                       Science, Sweden
Mauro Gaggero   National Research Council of Italy, Italy
Mauro Gaspari   Department of Computer Science and
                                       Engineering, University of Bologna, Italy
Bogdan Gliwa   AGH University of Science and Technology,
                                       Poland
Christos Grecos   Central Washington University, USA
Foteini Grivokostopoulou   University of Patras, Greece
Jerzy Grzymala-Busse   University of Kansas, USA
Ioannis Hatzilygeroudis   University of Patras, Greece
Enrique Herrera-Viedma   University of Granada, Spain
Dawn E. Holmes   University of California, Santa Barbara, USA
Katsuhiro Honda   Osaka Prefecture University, Japan
Chia-ling Hsu   TKU, Taiwan
Yuh-Jong Hu   National Chengchi University, Taipei, Taiwan
Yoshiteru Ishida   Toyohashi University of Technology, Japan
Naohiro Ishii   Aichi Institute of Technology, Japan
Mirjana Ivanovic   University of Novi Sad, Serbia
Yuji Iwahori   Chubu University, Japan
Lakhmi Jain   Bournemouth University, UK/University
                                       of Canberra, Australia
Nikita Jain   Rajasthan Technical University, India
Jacqueline Jarvis   Central Queensland University, Australia
Piotr Jedrzejowicz   Gdynia Maritime University, Poland
Björn Johansson   Lund University, Sweden
Nikos Karacapilidis   University of Patras, Greece
Pawel Kasprowski   Silesian University of Technology, Poland
Radoslaw Piotr Katarzyniak   Wrocław University of Science and Technology,
                                       Poland
Eizo Kinoshita   Meijo University, Japan
Frank Klawonn   Ostfalia University, Germany

Boris Kovalerchuk	Central Washington University, USA
Petia Koprinkova-Hristova	Bulgarian Academy of Sciences, Sofia
Marek Kretowski	Bialystok University of Technology, Poland
Dalia Kriksciuniene	Vilnius University, Lithuania
Vladimir Kurbalija	University of Novi Sad, Serbia
Kazuhiro Kuwabara	Ritsumeikan University, Japan
Chee Peng Lim	Deakin University, Australia
Pei-Chun Lin	Department of Information Engineering and Computer Science, Feng Chia University, Taiwan
Ivan Luković	University of Novi Sad, Serbia
Neel Mani	ADAPT Centre for Digital Content Technology, Dublin City University, Dublin, Ireland
Raimundas Matulevičius	University of Tartu, Estonia
Lyudmila Mihaylova	University of Sheffield, UK
Mohamed Arezki Mellal	M'Hamed Bougara University, Algeria
Michael Möhring	Munich University of Applied Sciences, Germany
Daniel Moldt	University of Hamburg, Germany
Stefania Montani	DISIT, Computer Science Institute, University of Piemonte Orientale, Alessandria, Italy
Mikhail Moshkov	King Abdullah University of Science and Technology, Saudi Arabia
Fionn Murtagh	University of Huddersfield, UK
Kazumi Nakamatsu	University of Hyogo, Japan
Tomoharu Nakashima	Osaka Prefecture University, Japan
Shastri L. Nimmagadda	Curtin Business School, Curtin University, Perth, Australia
Takahiro Niwa	Aichi Institute of Technology, Japan
Selmin Nurcan	University of Paris, France
Marek Ogiela	AGH University of Science and Technology, Krakow, Poland
Takao Ohya	Kokushikan University, Japan
Takeshi Okamoto	Kanagawa Institute of Technology, Japan
Tomasz Orzechowski	AGH University of Science and Technology, Krakow, Poland
Georg Peters	Munich University of Applied Sciences, Germany
Camelia Pintea	Technical University Cluj-Napoca, Romania
Clara Pizzuti	National Research Council of Italy (CNR), Italy
Petra Perner	Institute of Computer Vision and Applied Computer Sciences, Germany
Isidoros Perikos	University of Patras & TEI of Western Greece, Greece
Gunther Piller	University of Mainz, Germany

Radu-Emil Precup	Politehnica University of Timisoara, Romania
Jim Prentzas	Democritus University of Thrace, Greece
Małgorzata Przybyła-Kasperek	University of Silesia in Katowice, Poland
Seema Purohit	Kirti College, Mumbai, India
Marcos Quiles	Federal University of Sao Paulo (UNIFESP), Brazil
Milos Radovanovic	University of Novi Sad, Serbia
Hirosato Seki	Osaka University, Japan
Sheela Ramanna	University of Winnipeg, Canada
Torsten Reiners	Curtin Business School, Curtin University, Perth, Australia
Paolo Remagnino	Kingston University Surrey, UK
Ana Respicio	Universidade de Lisboa, Portugal
John Ronczka	SCOTTYNCC Independent Research Scientist, Australia
Mika Sato-Ilic	University of Tsukuba, Japan
Kurt Sandkuhl	University of Rostock, Germany
Miloš Savić	University of Novi Sad, Faculty of Sciences, Department of Mathematics and Informatics, Serbia
Rainer Schmidt	Munich University, Germany
Christian Schweda	Technical University of Munich, Germany
Ralf Seepold	HTWG Konstanz, Germany
Aleksander Skakovski	Gdynia Maritime University, Poland
Marina Sokolova	Southwest State University, Russia
Urszula Stańczyk	Silesian University of Technology, Gliwice, Poland
Ulrike Steffens	Hamburg University of Applied Sciences, Germany
Janis Stirna	Stockholm University, Sweden
Ruxandra Stoean	University of Craiova, Romania
Mika Sulkava	Natural Resources Institute Finland (Luke), Finland
Shing Chiang Tan	Multimedia University, Malaysia
Maasnori Takagi	Iwate Prefectural University, Japan
Dilhan J. Thilakarathne	VU University Amsterdam/ING Bank Personeel BV, The Netherlands
Ippei Torii	Aichi Institute of Technology, Japan
Jeffrey W. Tweedale	DST Group, Australia
Eiji Uchino	Yamaguchi University, Japan
Marco Vannucci	Scuola Superiore Sant'Anna, Italy
Pandian Vasant	Universiti Teknologi PETRONAS, Malaysia
Fen Wang	Central Washington University, USA
Junzo Watada	Universiti Teknologi PETRONAS, Malaysia

Yuji Watanabe	Nagoya City University, Japan
Ing Matthias Wissotzki	University of Rostock, Germany
Gloria Wren	Loyola University Maryland, USA
Celimuge Wu	The University of Electro-Communications, Tokyo, Japan
Yoshiyuki Yabuuchi	Shimonoseki City University, Japan
Takahira Yamaguchi	Keio University, Japan
Hiroyuki Yoshida	Harvard Medical School, USA
Beata Zielosko	University of Silesia in Katowice, Institute of Computer Science, Poland
Alfred Zimmermann	Reutlingen University, Germany
Sergey Zykov	National Research University Higher School of Economics and National Research University MEPhI, Russia

Contents

**Designing Interactive Topic Discovery Systems
for Research and Decision Making**........................... 1
Aneesha Bakharia

**Daily Stress Recognition System Using Activity Tracker
and Smartphone Based on Physical Activity
and Heart Rate Data** 11
Worawat Lawanont, Pornchai Mongkolnam, Chakarida Nukoolkit,
and Masahiro Inoue

**Listwise Recommendation Approach with Non-negative
Matrix Factorization** 22
Gaurav Pandey and Shuaiqiang Wang

**Estimation of Business Demography Statistics:
A Method for Analyzing Job Creation and Destruction** 33
Masao Takahashi, Mika Sato-Ilic, and Motoi Okamoto

Multivariate Multiple Orthogonal Linear Regression 44
Thi Binh An Duong, Jun Tsuchida, and Hiroshi Yadohisa

**Supervised Multiclass Classifier for Autocoding Based
on Partition Coefficient**..................................... 54
Yukako Toko, Kazumi Wada, Shinya Iijima, and Mika Sato-Ilic

**An Empirical Analysis of Volatility by the SIML Estimation
with High-Frequency Trades and Quotes**....................... 65
Hiroumi Misaki

**AFRYCA 3.0: An Improved Framework for Consensus
Analysis in Group Decision Making**........................... 76
Álvaro Labella and Luis Martínez

An Application of Transfer Learning for Maritime Vision
Processing Using Machine Learning   87
Jeffrey W. Tweedale

Fuzzy Regression Model Dealing with Vague Possibility
Grades and Its Characteristics   98
Yoshiyuki Yabuuchi

Decision-Oriented Composition Architecture for Digital
Transformation 109
Alfred Zimmermann, Rainer Schmidt, Kurt Sandkuhl, Dierk Jugel,
Justus Bogner, and Michael Möhring

Revenue Management Information Systems for Small
and Medium-Sized Hotels: Empirical Insights into
the Current Situation in Germany 120
Michael Möhring, Barbara Keller, Rainer Schmidt,
and Alfred Zimmermann

On Knowledge-Based Design Science Information System
(DSIS) for Managing the Unconventional Digital Petroleum
Ecosystems 128
Shastri L. Nimmagadda, Neel Mani, and Torsten Reiners

On a Smart Digital Gender Ecosystem for Managing
the Socio-Economic Development in the African Contexts 139
Christina Namugenyi, Shastri L. Nimmagadda, and Neel Mani

A Framework for a Dynamic Inter-connection of Collaborating
Agents with Multi-layered Application Abstraction
Based on a Software-Bus System 150
Robert Brehm, Mareike Redder, Gordon Flaegel, Jendrik Menz,
and Cecil Bruce-Boye

Non-reciprocal Pairwise Comparisons and Solution
Method in AHP 158
Kazutomo Nishizawa

Super Pairwise Comparison Matrix in the Multiple Dominant
AHP with Hierarchical Criteria 166
Takao Ohya and Eizo Kinoshita

A Questionnaire Method of Class Evaluations Using AHP
with a Ternary Graph 173
Natsumi Oyamaguchi, Hiroyuki Tajima, and Isamu Okada

A Link Diagram for Pairwise Comparisons 181
Takafumi Mizuno

**Measurement of Abnormality in Eye Movement with Autism
and Application for Detect Fatigue Level** 187
Ippei Torii, Takahito Niwa, and Naohiro Ishii

**Measurement of Line-of-Sight Detection Using Pixel Quantity
Variation for Autistic Disorder** 197
Takahito Niwa, Ippei Torii, and Naohiro Ishii

Irradiation of the Ear with Light of LED for Self-awakening 207
Mateus Yudi Ogikubo, Ippei Torii, and Naohiro Ishii

Projection Mapping of Animation by Object Recognition 217
Kazuya Yonemoto, Ippei Torii, and Naohiro Ishii

Development of Recovery of Eye-Fatigue by VDT Works 227
Ryotaro Kodera, Ippei Torii, and Naohiro Ishii

Author Index 237

Designing Interactive Topic Discovery Systems for Research and Decision Making

Aneesha Bakharia[✉]

Institute for Learning and Teaching Innovation, The University of Queensland,
Brisbane, QLD, Australia
aneesha.bakharia@gmail.com

Abstract. The introduction of interactive variants of Non-negative Matrix Factorization and Latent Dirichlet Allocation algorithms has made it possible to develop systems that support human in the loop interaction and refinement of topic models. This paper presents design guidelines for developing software that is able to interactively support researchers and decision makers in the discovery and interpretation of topics found in both small and large domain-specific document collections. The guidelines emerged from the analysis of two between-subjects experiments, which involved comparing algorithmically derived topics with the topics produced by research analysts and evaluating interactive topic modeling algorithms. A key finding from both experiments was that a holistic approach that includes additional algorithms needs to be taken when designing human in the loop systems to support research and decision making. The paper also outlines key areas where additional research is required in order to comply with the proposed design guidelines. Much of the proposed functionality is not currently included in topic model exploration software.

Keywords: Interactive topic modeling
Non-negative matrix factorization · Latent dirichlet allocation
Decision support · Knowledge management · Qualitative research
Qualitative content analysis

1 Introduction

Non-negative Matrix Factorization (NMF) [12] and Latent Dirichlet Allocation (LDA) [4] are two (2) popular and useful algorithms that are able to find latent topics within document collections. While NMF and LDA stem from different mathematical underpinnings, both algorithms are able to map documents to topics and words to topics. Figure 1 shows the top words and documents for a derived topic on "Beginning Teachers". NMF and LDA are also not hard clustering algorithms as they allow documents and words to be mapped to multiple topics, a feature that mimics the way, humans, cluster documents. Many users,

© Springer International Publishing AG, part of Springer Nature 2019
I. Czarnowski et al. (Eds.): KES-IDT 2018, SIST 97, pp. 1–10, 2019.
https://doi.org/10.1007/978-3-319-92028-3_1

particularly if they come from a research background, don't trust the output of topic modeling algorithms [2]. The main reasons for the lack of trust are that the algorithms are perceived to function as black boxes, are unable to support inter-activity and don't improve with human feedback. Recently, however, interactive variants for both NMF [14] and LDA [1] have been proposed that allow domain knowledge to be provided to the algorithm in the form of specifying topic seed words (i.e., to create a new topic from a list of words) and identifying topics that need to either be merged or split.

Interactive variants of NMF [14] and LDA [1] are capable of addressing user trust concerns. It is rare that when an algorithm is first introduced for the evaluation to include human participants. Topic modeling algorithms have been found to be useful in numerous domain applications but in order for their use to become mainstream particularly by research analysts (including qualitative content analysts and decision makers), additional research is required to under-stand how humans interpret topics and whether they are able to effectively use the built-in algorithmic mechanisms to add their domain knowledge to refine topics and answer research questions. While studies evaluating interactive topic modeling algorithms with human participants are starting to be published [3,6], this important field of research is in its infancy. In this paper, the results of two (2) experiments involving human participants were analyzed to produce design guidelines for the development of interactive topic discovery software. Eleven (11) design guidelines are included in this paper and provide insight into the functionality required to support human in the loop interaction with topic mod-eling algorithms.

2 Methodology

The research methodology was to carefully chosen to design participant experi-ments that addressed current flaws in research studies where topic models have been evaluated by humans. In many research studies, participants are only shown the top words in a topic and asked to evaluate the coherence of the topic. How-ever, when users need to answer research questions and use the clustered docu-ment grouping information to make decisions, the context of where the top words occurred (i.e., location within the document) in the textual documents is very important. The top words in a topic help users interpret topics and keyword-in-context functionality supports the evidence gathering process. The analysis of the output of a topic modeling algorithm is also an iterative process, with the user reviewing both the top words and the top documents in a topic, supplying their domain knowledge in the form of feedback to the algorithm and iteratively refining the derived topics. In most previous research studies, participants are also asked to evaluate topic coherence without being given a research task or question that they need to answer. In the two experiments described in this paper, the context of where top words appear in the documents that belong to a topic, the notion of iterative interactive topic refinement and the assignment of a research task to participants has been addressed. In this section, Experiment

1: Determining Interactivity Requirements and Experiment 2: Evaluating Interactive Topic Modeling Algorithms are described. Both experiments were used to derive design guidelines for interactive topic modeling software. It should be noted that this paper does not contain the detailed results of each experiment, and instead discusses the high-level features that participants found valuable and problematic. Each high-level issue is also addressed with a specific design guideline. The design guidelines are proposed to inform future interactive topic discovery systems designed for researchers and decision-makers.

2.1 Experiment 1: Determining Interactivity Requirements

In Experiment 1: Determining Interactivity Requirements a between subjects experiment was conducted with participants split into two (2) groups of 10 participants, to either manually analyze a dataset or evaluate the topics derived with the aid of a topic modeling algorithm. A small corpus consisting of teacher responses to a survey on professional development was selected as it contained multiple overlapping topics and was able to be analyzed manually in a short timeframe (i.e., approximately 1 h). Group A consisted of manual coders that were required to read the corpus and write down the main topics and the survey responses that belonged to the topic. Group B was provided with a simple web-based interface where they could specify the number of topics to be generated and then view the main words and responses (i.e., documents) that belonged to a topic. All user interactivity with the algorithm including changing the number of topics was tracked. Participants in Group B were required to rate the generated topics and complete a questionnaire to identify the types of interactivity that would be useful in addressing a research question. Based on the findings of Experiment 1, the Penalized NMF (PNMF) [14] and Dirichlet Forest LDA (DF-LDA) [1] algorithms were selected as the interactive variants to be evaluated in Experiment 2. PNMF incorporates constrained clustering ideas and includes a regularization penalty term that supports both must-link (i.e., topic seeding and topic merging) and cannot-link constraints (i.e., splitting). DF-LDA is an extension to the LDA probabilistic model that allows for the addition of domain knowledge via the specification of must-link and cannot-link sets of words.

2.2 Experiment 2: Evaluating Interactive Topic Modeling Algorithms

In Experiment 2: Evaluating Interactive Topic Modeling Algorithms, a between-subjects experiment was conducted to compare an NMF interactive variant with a comparable LDA interactive variant. Both groups of participants were presented with a user interface designed to support topic creation (i.e., the specification of words that must be grouped together in a topic) and the merging and splitting of topics. All participant interaction, including changing the number of topics to be generated and the time spent reviewing topics was tracked. The analysis of the results from Experiment 2 (including the analysis of all tracked interaction with the algorithms, the time spent reviewing the derived topics, the

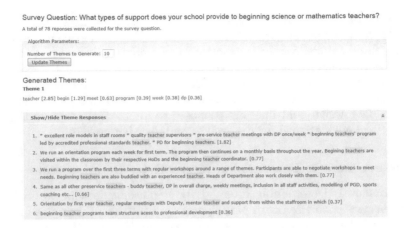

Fig. 1. The user interface to display derived topics for Group B participants in Experiment 1.

final rated topics, and the completed questionnaire) led to additional guidelines for the development of interactive topic modeling algorithms. Participants were provided with a research question to answer and asked to identify topics that contained evidence to support a research finding (Fig. 2).

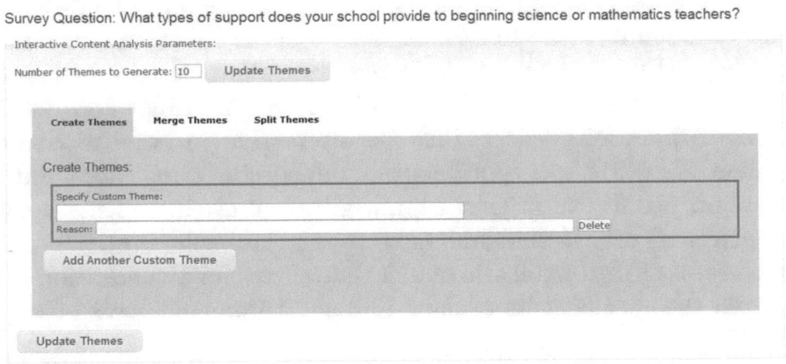

Fig. 2. The user interface designed to support the addition of domain knowledge in Experiment 2.

2.3   Corpus Preprocessing

The corpus preprocessing steps included stop word removal, word stemming, and the construction of a bag-of-words matrix (i.e., document-to-term matrix). The number of topics was included as a parameter that the participants had to interactively specify via the provided user interface. NMF variant algorithms were initialized with a Singular Value Decomposition technique [5] to ensure algorithm convergence.

3 Design Guideline Development

In this section design guidelines for interactive human in the loop topic discovery systems are proposed and substantiated with research findings from Experiment 1: Determining Interactivity Requirements and Experiment 2: Evaluating Interactive Topic Modeling Algorithms. The design guidelines support research and decision making processes.

In Experiment 1, participants completing the activity manually (Group A), frequently mapped a text response to two (2) or more topics thereby employing a soft-clustering approach. This suggests that algorithms that allow overlapping cluster membership are closely related to the way humans group documents together. NMF and LDA, in particular, are not hard clustering approaches. The importance of using algorithms that support overlap among topics justifies the first design guideline (Design Guideline 1). Widely used hard clustering algorithms such as k-means are inappropriate as interactive computational aids for topic discovery because they will not reveal overlap to the user.

> **Design Guideline 1:**
> Algorithms used as aids to interactive topic modeling need to discover and intuitively visualize topic overlap.

Text responses manually placed in finer grained topics were grouped together by participants (Group A) because they contained a phrase (i.e., multiple words) or where relationships existed between the words in the documents which were not directly reflected in the word usage between the documents. A useful form of interactivity would, therefore, take advantage of the user's domain knowledge and allow them to supply sets of related words or phrases or incorporate pre-built word embeddings [8].

> **Design Guideline 2:**
> Algorithms used as aids to interactive topic modeling need to allow the user to supply sets of related words as domain knowledge and incorporate pre-built word embeddings in order to produce topics that are similar to those that could be derived manually by research analysts.

In Experiment 1, participants that used a topic modeling algorithm (Group B) were not shown the whole corpus and had to specify the number of topics (i.e., k) that had to be generated using a simple user interface. Participants in Group B were then required to iteratively review and change the number of generated topics. Participants, in the post-task survey, said that it was difficult to select an appropriate value for k (i.e., the number of topics) because it was hard to retain the topics that were of a good quality across iterations. Recently, coherence metrics such as the UCI measure [10] and the UMass measure [9] have been proposed for the evaluation of topic modeling algorithms. These metrics have been shown to moderately correlate with human similarity judgments and could potentially provide a way to communicate topic quality to the user, thereby preventing researchers and decision makers from reviewing topics that are of poor quality. These metrics should be displayed with the derived topics to indicate topic quality, used as a way to rank topics and provide guidance on the selection of an initial value for k (Design Guideline 3). It should be noted that for the smaller and more domain-specific corpus that was used in Experiment 1 and 2, that both the UCI and UMass coherence measures were found not to correlate with the rating of topics by participants. In the paper introducing the UCI measure [10], a higher correlation (i.e., 0.7) was achieved for larger corpora with generic topics whereas for smaller domain-specific corpuses a lower correlation (i.e., 0.3) was documented. The impact topic overlap has on topic coherence has also not been evaluated. Additional research studies are therefore needed to develop coherence measures that match the human evaluation of overlapping topics derived from smaller domain-specific corpora.

Design Guideline 3:
Display metrics that indicate the topic coherence of a derived topic to help the user identify topic quality, rank topics and select an optimal number of topics.

The Experiment 1 user interface, displayed topics with their top words and text responses. A weighting was included for both the top words and documents in a topic. Participants were able to click on a word and view where the word occurred (i.e., keywords-in-context functionality). Numerous participants found it difficult to interpret the words in isolation and thought that multi-word expressions and phrases would be more useful. The data pre-processing separated the text responses into unigrams (i.e., single words for inclusion in the term-document matrix). An n-gram or equivalent approach that is able to identify phrases would be preferred by qualitative researchers and decision makers because multi-word expressions and phrases improve the interpretation of topics (Design Guideline 4). Additional algorithms can also be used such as a topic labeling technique [7].

Design Guideline 4:
Include phrases or multi-word expressions as features in the bag-of-words model to allow the top weighted phrases or multi-word expressions in a topic to be displayed in order to improve topic interpretation.

In Experiment 1, the only form of interactivity provided to participants in Group B was the ability to change the number of topics derived by the NMF algorithm (i.e., the k parameter). It was evident from the large and uneven fluctuations in values for k that participants found it difficult to locate the ideal number of topics in the corpus. The problem worsened in Experiment 2, where participants were able to specify topic seed words, as well as merge and split topics (see Fig. 3). The choice of an optimal k value (i.e., the number of topics) is parameter selection rather than a form of interactivity. Parameter selection and fine-tuning is a difficult task, even for machine learning experts. Due care must be taken in exposing algorithm parameters directly via a user interface and where possible other algorithms should be used to provide the user with a good initial value for k. This finding supports the inclusion of Design Guideline 5, which focuses on using additional algorithms to aid the analyst in finding a good approximation of the number of topics. An aggregate of individual topic coherence measures for each k value has been shown to produce a good initial starting value [11].

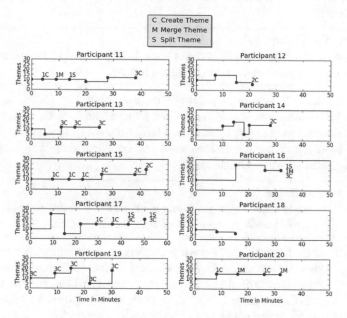

Fig. 3. Group A participants selection of k and the types of rules specified in Experiment 2.

Design Guideline 5:

Use parameter initialization techniques to aid users with parameter selection (e.g., specifying the number of topics) for topic modeling algorithms.

It was evident that the display of the top-weighted words in a topic played a role in helping participants to interpret topics and justifies the inclusion of Design Guideline 6. Displaying the raw scores from the term-topic matrix next to a word, however, was found to confuse participants. An alternate technique to express the raw word scores from NMF as a probability needs to be investigated.

Design Guideline 6:

Allow users to view the top weighted words from a topic as a means by which to interpret topics at a high-level.

Allowing participants to click on the top weighted words and view where they were used within the documents, assisted participants with topic interpretation. The inclusion of tools that facilitate in-context analysis such as keyword-in-context tools also led to improved topic interpretation (Design Guideline 7). The inclusion of keywords-in-context tools facilitated an enhanced understanding of the topic and provided validating evidence. Current tools that display the output of topic modeling algorithms [13] don't show the top documents in a topic or allow the user to view where the top words have appeared within text snippets. Findings from the two experiments conducted, found the inclusion of keyword-in-context functionality crucial to support topic interpretation and evidence gathering.

Design Guideline 7:
The inclusion of keyword-in-context tools is important to facilitate topic interpretation and support evidence gathering.

In Experiment 1, Group B participants answered the following free text question, "What other manipulation and/or interaction techniques can you suggest?". The results were analyzed and used to determine the interactivity requirements for topic modeling algorithms used in Experiment 2. The requirements are summarized in Design Guideline 8.

Design Guideline 8:
Interactive topic modeling algorithms need to allow users to specify new topics (i.e., topic seeding), merge topics and split topics.

In Experiment 2, participants were provided with a simple user interface to specify new topics, merge and split topics. Participants in Experiment 2, mainly used the ability to provide seed words (i.e., topic creation) to produce topics that were able to address the research question they were required to address in the study. Experienced participants (i.e., those with extensive qualitative research experience at a doctoral level) were able to specify acronyms, and sets of related words. Experienced participants were also able to provide the necessary topic merge and split rules. When merge and split rules were used, participants were able to obtain additional topics relevant to the research questions they were provided. Active Machine Learning techniques should be investigated as a means to bridge the gap between experienced and novice users by perhaps identifying topics that need to be merged or split and recommending seed words for topics based on distributional semantics. The recommendation of additional seed words for topics via the inclusion of pre-built Word2Vec embeddings [8] and Active Machine Learning algorithms requires further investigation.

Design Guideline 9:
Interactivity should not only be one-way between the user and the algorithm, models of interactivity based on Active Machine Learning and word embedding techniques must be leveraged to facilitate algorithm-user interactivity.

In Experiment 2, participants found it difficult to locate the topics that were the results of the seed words they entered, or the topics that were either merged or split because the order of topics produced by the algorithms was always different (i.e., both NMF and LDA are non-deterministic). Design Guideline 10 seeks to assist users in finding the topics that were derived from the addition of domain knowledge by using similarity metrics.

Design Guideline 10:
Use similarity metrics to find derived themes that have been influenced by the domain knowledge supplied by the user.

The user interface provided to participants in Experiment 2 was very simplistic. Careful thought and design must be applied to the construction of user interface elements to support the addition of domain knowledge to deliver an intuitive and easy to use interface. In particular, the process of adding merge and split rules needs to be scaffolded.

Design Guideline 11:
Design intuitive user interface elements that support the creation of create, merge and split rules for interactive topic modeling algorithms.

---

4   Conclusion and Future Directions

In this paper, the findings of two (2) experiments using both generic and interactive variants of NMF and LDA were used to derive design guidelines for interactive topic discovery software systems. The proposed guidelines are practical and indicate that a holistic approach to the design and implementation of topic discovery software are essential when humans need to answer specific research questions and make decisions based on their findings. In particular, the inclusion of numerous algorithms in addition to the interactive variants of NMF and LDA are required. Key additional algorithms include, algorithms to aid in the selection of the number of topics k, algorithms that are able to calculate topic coherence on domain-specific corpora, active learning algorithms to enable two-way initiated interactivity (i.e., human-algorithm and algorithm-human), and similarity metrics to help the user find topics that have emerged as a result of user-supplied domain knowledge (i.e., seed words to create topics, topic merge rules, and topic split rules). Intuitive user interface design is also required to assist the user in specifying domain knowledge. Future research directions include developing and evaluating an open source system that incorporates the design guidelines outlined in this paper.

Acknowledgements. The experiments described within this paper were conducted as part of my doctorate degree at Queensland University of Technology. I would like to thank and acknowledge my supervisors Peter Bruza, Jim Watters, Bhuva Narayan and Laurianne Sitbon.

References

1. Andrzejewski, D., Zhu, X., Craven, M.: Incorporating domain knowledge into topic modeling via dirichlet forest priors. In: Proceedings of the 26th Annual International Conference on Machine Learning, pp. 25–32. ACM (2009)
2. Bakharia, A.: Interactive content analysis: evaluating interactive variants of non-negative Matrix Factorisation and Latent Dirichlet Allocation as qualitative content analysis aids. Ph.D. thesis, Queensland University of Technology (2014)
3. Bakharia, A., Bruza, P., Watters, J., Narayan, B., Sitbon, L.: Interactive topic modeling for aiding qualitative content analysis. In: Proceedings of the 2016 ACM on Conference on Human Information Interaction and Retrieval, pp. 213–222. ACM (2016)
4. Blei, D.M., Ng, A.Y., Jordan, M.I.: Latent dirichlet allocation. J. Mach. Learn. Res. **3**, 993–1022 (2003)
5. Boutsidis, C., Gallopoulos, E.: SVD based initialization: a head start for nonnegative matrix factorization. Pattern Recogn. **41**(4), 1350–1362 (2008)
6. Lee, T.Y., Smith, A., Seppi, K., Elmqvist, N., Boyd-Graber, J., Findlater, L.: The human touch: how non-expert users perceive, interpret, and fix topic models. Int. J. Hum Comput Stud. **105**, 28–42 (2017)
7. Mei, Q., Shen, X., Zhai, C.: Automatic labeling of multinomial topic models. In: Proceedings of the 13th ACM SIGKDD International Conference on Knowledge Discovery and Data Mining, pp. 490–499. ACM (2007)
8. Mikolov, T., Sutskever, I., Chen, K., Corrado, G.S., Dean, J.: Distributed representations of words and phrases and their compositionality. In: Advances in Neural Information Processing Systems, pp. 3111–3119 (2013)
9. Mimno, D., Wallach, H.M., Talley, E., Leenders, M., McCallum, A.: Optimizing semantic coherence in topic models. In: Proceedings of the Conference on Empirical Methods in Natural Language Processing, pp. 262–272. Association for Computational Linguistics (2011)
10. Newman, D., Noh, Y., Talley, E., Karimi, S., Baldwin, T.: Evaluating topic models for digital libraries. In: Proceedings of the 10th Annual Joint Conference on Digital Libraries, pp. 215–224. ACM (2010)
11. O'Callaghan, D., Greene, D., Carthy, J., Cunningham, P.: An analysis of the coherence of descriptors in topic modeling. Expert Syst. Appl. **42**(13), 5645–5657 (2015)
12. Seung, D., Lee, L.: Algorithms for non-negative matrix factorization. Adv. Neural Inf. Proc. Syst. **13**, 556–562 (2001)
13. Sievert, C., Shirley, K.E.: LDAvis: A method for visualizing and interpreting topics (2014)
14. Wang, F., Li, T., Zhang, C.: Semi-supervised clustering via matrix factorization. In: SDM, pp. 1–12 (2008)

Daily Stress Recognition System Using Activity Tracker and Smartphone Based on Physical Activity and Heart Rate Data

Worawat Lawanont[1](✉), Pornchai Mongkolnam[2], Chakarida Nukoolkit[2], and Masahiro Inoue[1]

[1] Graduate School of Engineering and Science, Shibaura Institute of Technology, Saitama, Japan
nb16501@shibaura-it.ac.jp
[2] School of Information Technology, King Mongkut's University of Technology Thonburi, Bangkok, Thailand

Abstract. Everyday, people experience stress, and it has been suggested for a long time that stress will eventually develop into anxiety as well as other physical issues. The emerging technology, such as wearable sensors and smartphone, have enabled the opportunity of using the technology to help solve the issue. In this paper, we proposed a system using Internet of Things architecture where we adopted an activity tracker as our sensing device to reduce cumbersome for daily use. Among the total of 17 features extracted from activity tracker, five features from sleep data and six features from heart rate data were proposed to develop the stress recognition model. In the evaluation of our system, we achieved the accuracy as high as 81.70% on the cross validation and 78.95% when tested on the test set. Despite that this is a preliminary result, it has shown that it is possible to use the IoT architecture along with the activity tracker to accurately recognize stress and help improve one's wellbeing.

Keywords: Stress recognition · Activity tracker · Internet of Things
Digital healthcare · Wearable sensor

1 Introduction

Daily stress has been a topic for a long time. Despite there are many factors related to stress, the consequences of stress obviously affect the quality of life and wellbeing. A study suggested that physical symptoms are likely to occur after a stressful day [3]. In general, stress management is an important issue and it is necessary to make use of daily technology and devices to help improve one's wellbeing as much as possible.

In this paper, we proposed a system that is based on Internet of Things (IoT) architecture to recognize stress level in daily life. The system adopted an activity tracker as a sensing device. The activity tracker itself is also capable of sensing

© Springer International Publishing AG, part of Springer Nature 2019
I. Czarnowski et al. (Eds.): KES-IDT 2018, SIST 97, pp. 11–21, 2019.
https://doi.org/10.1007/978-3-319-92028-3_2

heart rate, which is an important feature for stress recognition [15]. The system was developed based on the idea that it must be practical for daily usage and it should work with devices that can be easily found in the market.

2   Related Work

After the emerging of technology, countless of studies were conducted to improve one's wellbeing. These studies include several perspectives, one is the study conducted in clinical setting to help the patient with chronic diseases or similar cases, while the other studies try to improve the wellbeing by using daily life technology to solve the matter [8,17]. The activity recognition was further used for stress recognition as well. Nicholas and Jean-Marc proposed a study to develop a supervised learning method for recognizing stress [11]. The study collected physiological data in a stressful situation that was built for the experiment. Beside from physiological data classification model, a study discussed stress recognition using smartphone data, weather conditions, and individual traits [1]. On the other hand, a study called StressSense used smartphone to recognize the stress [9]. These two studies demonstrated good examples of using a simple everyday device, such as smartphone, to help improve one's wellbeing.

As an extension to smartphone data, such as number of calls and physical activities, another parameter that was proposed to help recognize the stress was heart rate variability (HRV) [15]. The study findings suggested that the HRV may index important organism functions and it should be considered as a potential stress marker. With the use of heart rate approach, a study used a low cost chest strap heart rate sensor to measure stress [13]. Despite that the result was very satisfying, the heart rate chest strap is still not widely used. On another perspective, a study by Muaremi et al. showed an example of measuring stress using the combination of smartphone and wearable devices [12]. However, the wearable device used in the study is also not widely used.

In general, the approach of using smartphone and wearable devices were adopted for stress recognition. However, the works which were feasible for daily usage did not use heart rate or HRV as one of the parameter for model training. In addition, the user was required to wear a special device, such as chest strap, for heart rate monitoring. Thus, we proposed a solution where the system used smartphone and activity tracker for stress recognition while consider both physical activities data as well as heart rate data from wearable device.

3   Methodology

3.1   System Overview

Figure 1 shows the overview of the proposed system. This study designed the system according to the ETSI M2M architecture [6]. In this system, we divided the system into two main parts. The body area network consists of two main components, activity tracker and smartphone. The activity tracker serves as a sensor device that collects data from the user. The smartphone serves as a body area gateway. Its main task is to communicate and receive data from the activity tracker. Then, it processes them before sending them to the cloud service using the Internet connection. Moreover, as an extension to this system, the use of body area gateway also enables the possibility of multiple sensor device implementation. As each device may not has Internet connectivity, the body area gateway can help gather the data, organize the data, and prepare the data before sending to cloud service. This reduces the traffic and bandwidth that is require between the devices and cloud service.

For the cloud service, the first component, database, stores the data received from the body gateway. Moreover, the database stores personal information of the user and the classified data from the machine learning component. The machine learning component provides classification service, such as low or high stress level, and visualization to the user such that the visualization part provides feedback to user via the smartphone application.

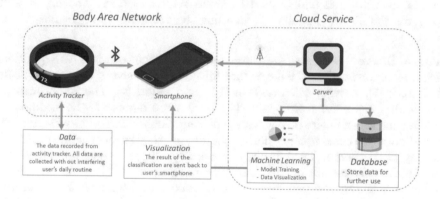

Fig. 1. Overview of the daily stress recognition system

3.2 Experiment Procedure

We conducted the experiment using a FitBit activity tracker. The tracker's heart rate sensor was proved to have no significant difference in sensing the heart rate compared to other famous trackers in the market [16]. First, all of the subjects were informed about the detail of the study as well as its purposes. Then, the subjects were asked to complete two surveys. The third survey was also given to them but they were asked to complete and return it the next day. Then, the data collection process started. After the data collection was completed on the next day, we prepared the data for model development, which was done when all data were received from every subject. Finally, we evaluated the model as discussed in Sect. 4.

3.3 Survey

This study used the perception survey as an evaluation method for the model. The surveys chosen for the study were to determine the overall stress level. The survey did not pointed out any stress from certain specific tragic event, such as loss of love ones, instead, it focused on overall perceived stress level.

Perceived Stress Scale. Perceived Stress Scale (PSS) [2] is one of the most widely used survey for measuring perception of stress. The survey consists of 10 questions, each one is a 5 point scale question (maximum score = 40). The higher score results in a higher stress perceived by the subject over the past month.

General Self-Efficacy Scale. General Self-Efficacy Scale (GSE) [14] is a 10 questions 4 point scale survey (maximum score = 40). The purpose of the survey is to measure the ability of a person in coping with problems and demands in the past month. It is widely used in stress related studies. The higher score represents the higher capability in handling the problems and demands.

General Survey. We developed the general survey that consisted of 5 questions. Each question related to the factor that affected stress levels, which were health [5], fatigue [7], nervous [4], workload [10], and mood [4]. The survey was a 5 point scale (maximum score = 25). The total score represents the likelihood of experiencing stress where higher score represents higher possibility of stress. The general survey was completed during the day of the data collection where the morning part was completed during lunch and evening part was completed after they finished their work on that day. The purpose of this survey was that we wanted to examine the perceived stress of each day. While the PSS and GSE provided a stress level for the past month, the general survey could help to understand the stress level of each specific day.

3.4 Data Collection

This study collected the data from a total of 10 subjects, including 6 males and 4 females aged between 20–26 years old. The subjects participated in the experiment were researchers and graduate students who worked between 6–10 hours per day. The subjects did not have any medical condition and they were not taking any kind of medicine during the experiment period. First, the instructor informed the subjects about the information of this data collection and experiment and gave them three surveys, as mentioned in Sect. 3.2. Then, the data collection process start at night, where all subjects wore the activity trackers before sleep, and for the whole day of the next day. We collected the data from the subjects on their working day, where each subject had at least 6 working hours on the day. In total, we had 10 days of data from 10 subjects. The five types off data collected were number of steps, calories, sleep cycle, heart rate, and resting heart rate. In addition, we performed normality test on the collected data with Kolmogorov-Smirnov test. The null hypothesis for the test was that each type of data was normally distributed. The test at 5% significance level on all five types of data showed that it failed to reject the null hypothesis, and that the data came from a standard normal distribution.

3.5 Features Extraction

From the total of 5 types of data, we extract a total of 17 features on 1-h interval of collected data. The sleep cycle data were attached to each instance as it was recorded overnight. Table 1 shows the 17 features.

Table 1. 17 Features extracted from the collected data

I. Number of Steps (3 Features)	II. Calories Usage (3 Features)	III. Sleep Cycle (5 Features)	IV. Heart Rate (4 Features)	V. Resting Heart Rate (2 Features)
• Mean	• Mean	• Deep sleep time	• Mean	• Resting heart rate
• Standard deviation	• Standard deviation	• REM sleep time	• Standard deviation	• Difference from heart rate mean (percentage)
• Summation	• Summation	• Light sleep time	• Maximum	
		• Awake time	• Minimum	
		• Total sleep time		

4 Evaluation and Results

4.1 Survey Results

Every subject completed all three surveys and Table 2 shows the descriptive statistics of the three surveys' results. For the general survey, we calculated the mean from the two parts of the survey, which were morning and evening, and used it as the score. Please note that for GSE result, the higher score represented higher capability of handling problems and demands, which resulted in less stress. The maximum possible score of the surveys were 40, 40, and 25 respectively for PSS, GSE, and General Survey.

Table 2. Descriptive statistics of survey score

Survey	Mean	Standard deviation	Maximum	Minimum
PSS	18.60	4.81	25.00	21.00
GSE	28.80	3.97	37.00	22.00
General survey	10.45	2.20	14.00	7.00

From the results, we used the mean value of each survey as a separation mark to label the data instances from each subject. For PSS and general survey, the subjects with score above the mean of each survey were labeled as 'HIGH' and those with score lower than the mean were labeled as 'LOW' For GSE, the subjects with score higher than the mean were labeled as 'LOW' and those with score lower than the mean were labeled as 'HIGH'. Lastly, the combined label was based on the majority from the results of the three mentioned surveys. Please note that 'HIGH' represents higher risk of stress based one each survey type and 'LOW' represents lower risk of stress. Table 3 shows the survey score and labeled class of each subject.

Table 3. Survey score and label of all subjects for each survey type

Subject	PSS	GSE	General survey	PSS Label	GSE label	General survey label	Combined label
1	18	27	11.00	Low	High	Low	Low
2	10	37	10.00	Low	Low	Low	Low
3	24	22	10.5	High	High	High	High
4	25	26	14.00	High	High	High	High
5	18	27	7.00	Low	High	Low	Low
6	19	28	12.50	High	High	High	High
7	24	29	13.00	High	Low	High	High
8	14	32	8.00	Low	Low	Low	Low
9	19	30	10.50	High	Low	High	High
10	15	30	10.00	Low	Low	Low	Low

4.2 Model Training

We trained the model using three algorithms, including K-Nearest Neighbor (K = 5), Support Vector Machine (SVM), and Decision Tree. The extracted features were preprocessed by mean normalization technique. Thus, all features were in the same scale. The data from each subject were randomly separated into training set and test set with the ratio of 80% to 20%. There were a total of 105 records in the training set and 25 records in the test set. First, we performed the training process using 5-fold cross validation on the training set. We trained the models to predict the stress using PSS, GSE, general survey, and combined result as the responses. Then, we performed training task using principal component analysis (PCA) on the same training set, classes, and algorithms. In total, we have trained 24 models, including 6 models for each response (PSS, GSE, general survey, and combined result). Figure 2 shows the modeling process with 4 class types.

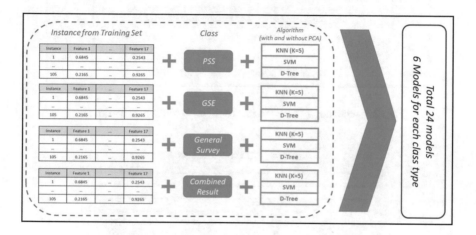

Fig. 2. Model training process with 4 class types

4.3   Cross Validation and Test Set

During the training process, we used K-fold cross validation (K = 5) as a model validation technique. Figure 3 shows the results of the model from the training. The highest accuracy when trained using PSS as a class was 85.40% with decision tree algorithm without PCA. The highest accuracy for using GSE as a class was 96.30% with SVM algorithm without PCA. For using the general survey as a class, the highest achieved accuracy was 81.70% with SVM algorithm without PCA. Lastly, when the model was trained with a combined survey results, the highest accuracy was 84.10% with decision tree algorithm without PCA.

Then, we selected the better performer model of each algorithm, which was that without PCA, and applied them to the test set. In this process, the model predicted the test set based on its own class type. Figure 4 shows the accuracy of each model when predicting the test set of its class. In all class types, model with Decision Tree performed better with the accuracy of all models higher than 78.00%. For the model predicting the combined survey result, the accuracy was 78.95%. The worst performing model was a KNN model used to predict the stress based on PSS result, where the accuracy was at 57.90%.

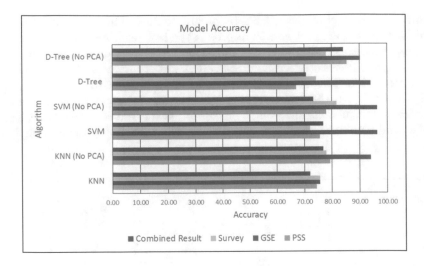

Fig. 3. Accuracy of the model from the 5-fold cross validation

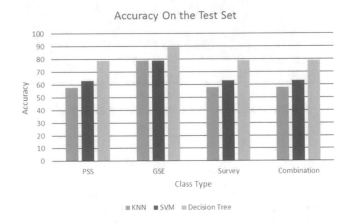

Fig. 4. Accuracy of the models when applied on the test set

5   Discussion

Overall, we trained the total of 24 models, including 6 models for predicting each class type (PSS, GSE, general survey and combined result). From the results of all the models we trained, the accuracies were in a satisfactory level as the best model (decision tree without PCA) achieved the accuracy of 84.10% when using the proposed combination of survey results as the class during the validation process. The highest accuracy when applying the model on the test set to predict the stress using the combination of survey results was 78.95%. Considering that both PSS and GSE results are used for one month of stress evaluation, the proposed classification method using combination of survey result as a response to the predictor was expected to provide more insight of daily stress instead of monthly stress. By being able to classify this stress level in daily manner, it will help users to understand there behavior better and encourage them to have a better routine activities for better wellbeing.

Moreover, the proposed features contributed very well to the model. The models that were trained with all the features were performing better than those trained with PCA in both cross validation and test set results. Furthermore, in this study we pointed out that heart rate and sleep data should be considered as one of the feature in classifying stress level and that the method in measuring the heart rate should not be difficult. We calculated the Pearson's correlation between the features and the combined label. The 5 features with the highest Pearson's correlation value were REM sleep time (0.5131), total sleep time (0.4493), resting heart rate (0.3984), deep sleep time (0.3373), and average heart rate (0.3010). In overall, we collected the data from the device that was feasible in daily use, and we created a model with the proposed features, which were a combination from multiple study.

Despite having promising results, there were also limitations. First, we noticed that the general survey approach was developed based on the factor related to stress. However, we believe that it can be improved to a more accurate survey in measuring daily perceived stress level. Secondly, the subjects and data set of this study were limited. A larger group of subjects with a longer monitoring and data collection period would provide a better insight on the classification model. It is also should be noted that this study did not conduct the experiment to measure the crosstalk level and accuracy of the device for each surrounding environment as the study was focusing on the model training process for classification. Lastly, the approach to develop a user-specific classification model is likely to work best in the long term. A system should slowly collect the data and retrain the model based on each user data for a personalized system.

6 Conclusion

In this paper, we proposed a solution for recognizing stress level using activity tracker and smartphone. The system was based on IoT architecture; therefore, it allowed further implementation of other sensing devices. We had developed the classification model using the proposed solution and the idea of using heart rate as a feature in recognizing the stress level. The study demonstrated the solution that was more practical for daily usage while maintaining the satisfactory accuracy level when compared to related study. Despite the fact that there are limitations to the study, there are also opportunities for further developments. This study contributes to modern lifestyle and shows the possibility of using daily electronic devices to encourage an individual to have better wellbeing.

Acknowledgments. This work was supported by JSPS KAKENHI Grant number 15K00929.

References

1. Bogomolov, A., Lepri, B., Ferron, M., Pianesi, F., Pentland, A.S.: Daily stress recognition from mobile phone data, weather conditions and individual traits. In: Proceedings of the 22nd ACM international conference on Multimedia, pp. 477–486. ACM (2014)
2. Cohen, S., Kamarck, T., Mermelstein, R., et al.: Perceived stress scale. Measuring stress: A guide for health and social scientists (1994)
3. DeLongis, A., Folkman, S., Lazarus, R.S.: The impact of daily stress on health and mood: psychological and social resources as mediators. J. Pers. Soc. Psychol. **54**(3), 486 (1988)
4. Frison, E., Eggermont, S.: The impact of daily stress on adolescents depressed mood: the role of social support seeking through Facebook. Comput. Hum. Behav. **44**, 315–325 (2015)
5. Galego, J.C.B., Moraes, A.M., Cordeiro, J.A., Tognola, W.A.: Chronic daily headache: stress and impact on the quality of life. Arq. Neuropsiquiatr. **65**(4B), 1126–1129 (2007)
6. Group, E., et al.: Open Service Access (OSA); Application Programming Interface (API); part 3: Framework (parlay5); etsies 203 915-3 v1. 2.1. ETSI/The Parlay Group (2007)
7. Hockey, R.: Stress and Fatigue in Human Performance, vol. 3. Wiley, New York (1983)
8. Lin, M., Lane, N.D., Mohammod, M., Yang, X., Lu, H., Cardone, G., Ali, S., Doryab, A., Berke, E., Campbell, A.T., et al.: BeWell+: multi-dimensional well-being monitoring with community-guided user feedback and energy optimization. In: Proceedings of the Conference on Wireless Health, p. 10. ACM (2012)
9. Lu, H., Frauendorfer, D., Rabbi, M., Mast, M.S., Chittaranjan, G.T., Campbell, A.T., Gatica-Perez, D., Choudhury, T.: Stresssense: detecting stress in unconstrained acoustic environments using smartphones. In: Proceedings of the 2012 ACM Conference on Ubiquitous Computing, pp. 351–360. ACM (2012)
10. Mansour, S., Mansour, S., Tremblay, D.G., Tremblay, D.G.: Workload, generic and work-family specific social supports and job stress: mediating role of work-family and family-work conflict. Int. J. Contemp. Hospitality Manage. **28**(8), 1778–1804 (2016)
11. Martin, N., Diverrez, J.M.: From physiological measures to an automatic recognition system of stress. In: International Conference on Human-Computer Interaction, pp. 172–176. Springer (2016)
12. Muaremi, A., Arnrich, B., Tröster, G.: Towards measuring stress with smartphones and wearable devices during workday and sleep. BioNanoScience **3**(2), 172–183 (2013)
13. Salai, M., Vassányi, I., Kósa, I.: Stress detection using low cost heart rate sensors. J. Healthc. Eng. 2016 (2016). Article ID 5136705
14. Schwarzer, R., Jerusalem, M.: The General Self-efficacy Scale (GSE). In: Weinman, J., Wright, S., Johnston, M. (eds.) Measures in Health Psychology: A Users Portfolio. Causal and Control Beliefs, pp. 35–37. NFER-NELSON (1995)
15. Thayer, J.F., Åhs, F., Fredrikson, M., Sollers, J.J., Wager, T.D.: A meta-analysis of heart rate variability and neuroimaging studies: implications for heart rate variability as a marker of stress and health. Neurosci. Biobehav. Rev. **36**(2), 747–756 (2012)

16. Wallen, M.P., Gomersall, S.R., Keating, S.E., Wisløff, U., Coombes, J.S.: Accuracy of heart rate watches: implications for weight management. PLoS ONE **11**(5), e0154420 (2016)
17. Weiss, G.M., Lockhart, J.W., Pulickal, T.T., McHugh, P.T., Ronan, I.H., Timko, J.L.: Actitracker: a smartphone-based activity recognition system for improving health and well-being. In: 2016 IEEE International Conference on Data Science and Advanced Analytics (DSAA), pp. 682–688. IEEE (2016)

Listwise Recommendation Approach with Non-negative Matrix Factorization

Gaurav Pandey[1(✉)] and Shuaiqiang Wang[2]

[1] University of Jyvaskyla, Jyväskylä, Finland
gaurav.g.pandey@jyu.fi
[2] Data Science Lab, jd.com, Beijing, China

Abstract. Matrix factorization (MF) is one of the most effective categories of recommendation algorithms, which makes predictions based on the user-item *rating matrix*. Nowadays many studies reveal that the ultimate goal of recommendations is to predict correct rankings of these unrated items. However, most of the pioneering efforts on *ranking-oriented* MF predict users' item ranking based on the original rating matrix, which fails to explicitly present users' preference ranking on items and thus might result in some accuracy loss. In this paper, we formulate a novel listwise user-ranking probability prediction problem for recommendations, that aims to utilize a *user-ranking probability matrix* to predict users' possible rankings on all items. For this, we present *LwRec*, a novel listwise ranking-oriented matrix factorization algorithm. It aims to predict the missing values in the user-ranking probability matrix, aiming that each row of the final predicted matrix should have a probability distribution similar to the original one. Extensive offline experiments on two benchmark datasets against several state-of-the-art baselines demonstrate the effectiveness of our proposal.

Keywords: Recommender systems · Collaborative filtering · Ranking

1 Introduction

Conventional recommendation algorithms like collaborative filtering follow a rating-oriented paradigm. They generally learn a recommendation model with users' observed historical ratings, using which they predict users' ratings on their unrated items. Nowadays, ranking-oriented recommender systems are receiving increasing attention from both academic communities and industry. Many studies reveal that the ultimate goal of recommendations is to predict correct rankings of these unrated items, and prediction of accurate ranking is more important than predicting accurate rating scores [1,5]. Accurate prediction of ratings does not necessarily imply improvement in the ranking results.

To elaborate, let us consider items: $\{a, b, c\}$ with their correct ratings $R :$ $\{5, 4, 3\}$ and two rating predictions $P_1 : \{3, 4, 5\}$ and $P_2 : \{3, 2, 1\}$. A rating oriented approach would prefer P_1 over P_2, since predicted ratings in P_1 are more

© Springer International Publishing AG, part of Springer Nature 2019
I. Czarnowski et al. (Eds.): KES-IDT 2018, SIST 97, pp. 22–32, 2019.
https://doi.org/10.1007/978-3-319-92028-3_3

accurate, being closer to the ratings in R. However, the order in P_1 $(a < b < c)$ is completely opposite to the desired order $(a > b > c)$. In contrast, a ranking oriented approach would prefer P_2, as it predicts the correct ranking, i.e. $(a > b > c)$. This would generate desirable results (correct order of items), in spite of having lower accuracy in predicted ratings.

Given the above argument, some pioneering efforts on *ranking-oriented* recommendation algorithms have been proposed. Due to the effectiveness of matrix factorization (MF) algorithms in rating-oriented recommender systems, a few ranking-oriented MF algorithms have been presented, reporting state-of-the-art results. However, most of the pioneering efforts on *ranking-oriented* MF predict users' item ranking based on rating scores, failing to explicitly present users' preference ranking on items and thus possibly resulting in some accuracy loss.

Therefore, we define a novel listwise user-ranking probability prediction problem for recommendations. We utilize the *listwise user-ranking probability matrix* [7] to explicitly characterize users' preference on items. Given a set of rating scores on items, each ranking on items might be possible, where "correct" rankings (higher scores are ranked at top positions) receive greater probabilities. Thus for each user, the probabilities on users' all possible rankings could formulate as a user-ranking probability matrix, where each element presents a probability that certain user holds certain ranking on items. Thus each row of the user-ranking probability matrix consists of users' probabilities on different item rankings, forming a distribution. With the initial probabilities of users' possible rankings on *their rated items*, the listwise user-ranking probability prediction problem aims to predict probabilities of users' possible rankings on *all items*. Meanwhile, the predictions should satisfy the requirements of probability distributions: each element in the probability matrix should be between 0 and 1, and sum of each row should be 1.

Given a collection of items, there might be a very large number of possible rankings (i.e. $n!$ rankings for n items), resulting in a extremely huge ranking probability matrix and calculations in training. In this study, we only consider the top-k ranked items in rankings, and the size of the matrix could be shrinked significantly, especially the size of the ranking probability matrix is equal to that of the user-item rating matrix when $k = 1$. Based on this matrix, we then present *LwRec*, a novel listwise ranking-oriented MF algorithm, which minimizes the difference between the initial distribution on the known rankings and the final distribution on all items with predictions for each user. Considering the non-negative property for each element, we adapt non-negative MF to implement *LwRec*. Our experimental results on benchmark datasets demonstrate significant performance gains over state-of-art recommender algorithms.

To summarize, our contributions are as follows. (1) We define a novel listwise user-ranking probability prediction problem for recommendations. (2) We present an effective algorithm to solve the problem based on non-negative MF. (3) We achieve significant performance gains against state-of-the-art recommendation algorithms on benchmark datasets.

The rest of the paper is organized as follows. Section 2 briefly presents the related work and Sect. 3 describes the problem formulation. Then, we explain the *LwRec* approach in Sect. 4 followed by experimental setup in Sect. 5. Finally, Sect. 6 presents the results and Sect. 7 concludes the paper.

2   Related Work

This section presents related work for collaborative filtering (CF) recommendation algorithms, which use only the ratings given by the users for the items, and do not need the domain knowledge. They are mainly of two types: rating oriented and ranking oriented. While rating oriented algorithms predict unknown item ratings for each user, ranking oriented algorithms predict item rankings. Both of them can be further categorized as memory-based or model based.

Rating Oriented Algorithms: Memory-based rating oriented algorithms are either user-based CF [6], that utilize similarities between users on the basis of available ratings; or item-based CF [14], that utilize similarities between items. Various advanced versions of this approach have been introduced. For example, SLIM [17] directly learns from the data, a sparse matrix of aggregation coefficients that are analogous to the traditional item-item similarities. FISM [9] learns the item-item similarity matrix as a product of two low-dimensional latent factor matrices. Model-based rating oriented algorithms aim to predict ratings by learning a model from observed ratings. Traditional model of this type is matrix factorization (MF) [10], that uses dimensionality reduction to decrease the distance between predicted and observed rating matrices. Some of the models that are based on matrix factorization are: Probabilistic MF [20], Non-negative MF [12], Factorization Machines [19], Hierarchical Poisson MF [4] and LLORMA [13].

Ranking Oriented Algorithms: EigenRank [15] is a well known ranking oriented memory based CF algorithm that follows the pairwise approach. It employs a greedy aggregation method to aggregate predicted pairwise preferences of items into total ranking. VSRank [23] represents users' pairwise preferences for items by using vector space model and utilizes the relative importances of each pairwise preference. Moreover, various model-based ranking oriented CF algorithms have been introduced that try to optimize a ranking oriented objective function. Some of the notable algorithms of this type are: CLiMF [21], CoFiRank [24], ListCF [7] and GBPR [18].

3   Problem Formulation

3.1   User-Ranking Probability Matrix

Considering m users and n items, for each user there are obviously $n!$ possible rankings of items. Given a set of rating scores on items, each ranking on items might be possible, where "correct" rankings (higher scores are ranked at top positions) receive greater probabilities. The probability of item rankings could be derived with the Plackett-Luce model [16], which is a widely used permutation (each permutation is actually a ranking) probability model in various domains. Each ranking ρ can be represented as an ordered list $(\rho_1, \rho_2 \ldots \rho_n)$, where ρ_i represents the item at the ith position, and positions of the items are unique. Hence, the probability of the ranking ρ can be calculated as:

$$Prob(\rho) = \prod_{i=1}^{n} \frac{\gamma(r_{\rho_i})}{\sum_{j=i}^{n} \gamma(r_{\rho_j})}, \tag{1}$$

where r_{ρ_i} is the rating for the item ρ_i and $\gamma(r) = e^r$.

Since, there are $n!$ rankings of items, which is a large number of rankings even for a small value of n, it makes the computation impractical. Hence, we employ the same approach as Huang et al. [7], that uses an alternative efficient method introduced by Cao et al. [3]. The approach focuses only on top k items in the rankings, leading to $\frac{n!}{(n-k)!}$ different top k sets. So, the probability of the rankings ρ_S whose top-k items are exact $S = \{i_1, i_2 \ldots i_k\}$ can be calculated as:

$$Prob(\rho_S) = \prod_{j=1}^{k} \frac{\gamma(r_{i_j})}{\sum_{l=j}^{n} \gamma(r_{i_l})} \tag{2}$$

We have m users and for each user we have $p = \frac{n!}{(n-k)!}$ ranking sets for top k items. Now, we construct the user-ranking probability matrix $\Theta_{m \times p}$. In Θ, each row corresponds to a particular user and contains the probabilities for the p rankings. To clarify, if $Prob_{u_i}(S_j)$ represents the probability of ranking S_j calculated for the user u_i (where $1 \leq j \leq p$ and $1 \leq i \leq m$), then:

$$\begin{aligned} \Theta_{i,j} = & Prob_{u_i}(S_j), \text{if ratings of all items in } S_j \text{ are known,} \\ & \perp, \text{otherwise} \end{aligned} \tag{3}$$

Especially when $k = 1$, i.e. when we consider only the top-1 items in all rankings, the size of Θ is equal to that of the user-item rating matrix. This is because, in this case, $p = \frac{n!}{(n-1)!} = n$.

3.2   Objective and Constraints

Given the matrix of known top k probabilities of items: $\Theta_{m \times p}$, where $p = \frac{n!}{(n-k)!}$, we aim to predict the unknown probabilities, that in turn can be used to generate recommendations. This can be achieved by using a listwise loss function and optimizing it using matrix factorization. For this, we define the following objective and the two related constraints:

Objective: Using two matrices $U_{z \times m}$ and $G_{z \times p}$ that construct the predicted probability matrix $U^\top G$, utilize a listwise loss function and matrix factorization to minimize the distance between Θ and $U^\top G$.

C1: Values in $U^\top G$ should be in the range 0 to 1 (as they are probabilities). i.e. $0 \leq U_{ij} \leq 1, \forall i = 1 \ldots z$ and $\forall j = 1 \ldots m$.

C2: Sum of each row of $U^\top G$ should be 1 (as a row contains probabilities of rankings for a particular user, that should sum up to 1). i.e. $\sum_{j=1}^{p} (U^\top G)_{ij} = 1, \forall i = 1 \ldots m$.

Definition 1 (Listwise User-Ranking Probability Prediction). *Given a user-ranking probability matrix Θ, where each observed element of $\Theta_{i,p}$ indicates certain user's probability for her certain (top-k) preference ranking on her rated items, and each row of Θ forms a probability distribution. The listwise user-ranking probability prediction problem aims to predict each user's probability of her top-k preference ranking on all items, where each row of $U^\top G$ forms a probability distribution as well after prediction, and each user's two distributions, observed and predicted, should be as similar as possible. Formally,*

$$\underset{U,G}{\arg\min} \sum_{i=1}^{m} diff(\Theta_i, (U^\top G)_i),$$

$$\text{s. t. } 0 \leq (U^\top G)_{ij} \leq 1, i = 1, 2, \ldots, m, \text{ and } j = 1, 2, \ldots, p \text{ (C1)} \qquad (4)$$

$$\sum_{j=1}^{p} (U^\top G)_{ij} = 1, i = 1, 2, \ldots, m \text{ (C2)}$$

Here $diff(\Theta_i, (U^\top G)_i)$ is the difference between two distributions Θ_i and $(U^\top G)_i$, i.e. the ith row of the user-ranking probability matrix before and after prediction.

4   Prediction Method

In this section, we present *LwRec* to solve our listwise user-ranking probability prediction problem. We use Kullback-Leibler divergence [11], a commonly used measure for calculating difference between probability distributions, to compute $diff(\Theta_i, (U^\top G)_i)$. In *LwRec*, we utilize non-negative matrix factorization (MF) [12] to implement our proposed algorithm, which can generate non-negative elements for U and G. Thus the elements of the user-ranking probability matrix $U^\top G$ are all non-negative. In order to satisfy the constraint C2, we introduce a collection of Lagrange penalty terms in the objective function $\sum_{j=1}^{p} (U^\top G)_{ij} = 1$ where $i = 1 \ldots m$. In standard Lagrange methods, the coefficients of Lagrange penalty terms could be either positive or negative, but in non-negative MF, all of the parameters have to be non-negative. Thus we introduce two non-negative vectors α and β, and regard $(\alpha_i - \beta_i)$ that can be either positive or negative, as the coefficient of the ith Lagrange penalty term to formulate our loss function. Moreover, addressing constraint C2 together with ensuring that the values in

$U^\top G$ are non-negative, also satisfies constraint C1 (i.e. values in $U^\top G$ should be in range 0 to 1). The formulation of our loss function can be presented formally as follows:

$$
\begin{aligned}
L(U, G, \alpha, \beta) = &-\sum_{i=1}^{m} \sum_{j=1, \Theta_{ij} \neq \perp}^{p} \Theta_{ij} \log \frac{(U^\top G)_{ij}}{\sum_{l=1, \Theta_{il} \neq \perp}^{p} (U^\top G)_{il}} \\
&+ \sum_{i=1}^{m} (\alpha_i - \beta_i)\left(\sum_{j=1}^{p} (U^\top G)_{ij} - 1 \right) + \frac{\lambda_1}{2}\|U\|^2 + \frac{\lambda_2}{2}\|G\|^2,
\end{aligned}
\tag{5}
$$

In Eq. 5, the first term represents the main optimization objective from Eq. 4, which defines the divergence between $U^\top G$ and observed probability matrix Θ. The second term is the weighted cumulative Lagrange penalty term for constraint C2. The last two terms are l2-norms of U and G to avoid over-fitting, where λ_1 and λ_2 are the respective coefficients. By expanding the log in $L(U, G, \alpha, \beta)$ and considering that the sum of each row in Θ is 1, the function can be reformulated as:

$$
\begin{aligned}
L(U, G, \alpha, \beta) = &\sum_{i=1}^{m} \log\left(\sum_{l=1, \Theta_{il} \neq \perp}^{p} (U^\top G)_{il} \right) - \sum_{i=1}^{m} \sum_{j=1, \Theta_{ij} \neq \perp}^{p} \Theta_{ij} \log \left((U^\top G)_{ij}\right) \\
&+ \sum_{i=1}^{m} (\alpha_i - \beta_i)\left(\sum_{j=1}^{p} (U^\top G)_{ij} - 1 \right) + \frac{\lambda_1}{2}\|U\|^2 + \frac{\lambda_2}{2}\|G\|^2
\end{aligned}
\tag{6}
$$

To minimize the loss function using gradient descent, we compute its gradients with respect to the variables U, G, α and β and derive the following updates:

$$
\begin{aligned}
U_{ia} \leftarrow U_{ia} - \eta_u &\Bigg(\frac{\sum_{l=1, \Theta_{il} \neq \perp}^{p} G_{al}}{\sum_{l=1, \Theta_{il} \neq \perp}^{p} (U^\top G)_{il}} - \sum_{j=1, \Theta_{ij} \neq \perp}^{p} \frac{\Theta_{ij} G_{aj}}{(U^\top G)_{ij}} \\
&+ (\alpha_i - \beta_i) \sum_{j=1}^{p} G_{aj} + \lambda_1 U_{ia} \Bigg), \\
G_{aj} \leftarrow G_{aj} - \eta_g &\Bigg(\sum_{i=1}^{m} \frac{U_{ia}}{\sum_{l=1, \Theta_{il} \neq \perp}^{p} (U^\top G)_{il}} - \sum_{i=1, \Theta_{ij} \neq \perp}^{m} \frac{\Theta_{ij} U_{ia}}{(U^\top G)_{ij}} \\
&+ \sum_{i=1}^{m} (\alpha_i - \beta_i) U_{ia} + \lambda_2 G_{aj} \Bigg), \\
\alpha_i \leftarrow \alpha_i - \eta_\alpha &\left(\sum_{j=1}^{p} (U^\top G)_{ij} - 1 \right) \text{ and } \beta_i \leftarrow \beta_i - \eta_\beta \left(1 - \sum_{j=1}^{p} (U^\top G)_{ij} \right)
\end{aligned}
\tag{7}
$$

where, η_u, η_g, η_α and η_β are the step sizes. Now, using non-negative matrix factorization [12], we choose the step sizes such that:

$$\eta_u = \frac{U_{ia}}{\frac{\sum_{l=1,\Theta_{il}\neq\perp}^{p} G_{al}}{\sum_{l=1,\Theta_{il}\neq\perp}^{p} (U^\top G)_{il}} + \alpha_i \sum_{j=1}^{p} G_{aj} + \lambda_1 U_{ia}},$$

$$\eta_g = \frac{G_{aj}}{\sum_{i=1}^{m} \frac{U_{ia}}{\sum_{l=1,\Theta_{il}\neq\perp}^{p} (U^\top G)_{il}} + \sum_{i=1}^{m} \alpha_i U_{ia} + \lambda_2 G_{aj}}, \quad (8)$$

$$\eta_\alpha = \frac{\alpha_i}{\sum_{j=1}^{p} (U^\top G)_{ij}} \text{ and } \eta_\beta = \beta_i$$

Substituting these values of the steps in updation formulas in Eq. 7, we derive the following multiplicative updates:

$$U_{ia} \leftarrow U_{ia} \frac{\sum_{j=1,\Theta_{ij}\neq\perp}^{p} \frac{\Theta_{ij} G_{aj}}{(U^\top G)_{ij}} + \beta_i \sum_{j=1}^{p} G_{aj}}{\frac{\sum_{l=1,\Theta_{il}\neq\perp}^{p} G_{al}}{\sum_{l=1,\Theta_{il}\neq\perp}^{p} (U^\top G)_{il}} + \alpha_i \sum_{j=1}^{p} G_{aj} + \lambda_1 U_{ia}},$$

$$G_{aj} \leftarrow G_{aj} \frac{\sum_{i=1,\Theta_{ij}\neq\perp}^{m} \frac{\Theta_{ij} U_{ia}}{(U^\top G)_{ij}} + \sum_{i=1}^{m} \beta_i U_{ia}}{\sum_{i=1}^{m} \frac{U_{ia}}{\sum_{l=1,\Theta_{il}\neq\perp}^{p} (U^\top G)_{il}} + \sum_{i=1}^{m} \alpha_i U_{ia} + +\lambda_2 G_{aj}}, \quad (9)$$

$$\alpha_i \leftarrow \frac{\alpha_i}{\sum_{j=1}^{p} (U^\top G)_{ij}} \text{ and } \beta_i \leftarrow \beta_i \sum_{j=1}^{p} (U^\top G)_{ij}$$

On optimizing U and G, The rows of $U^\top G$ would contain predicted probability distributions of top k rankings for users, that can be utilized to generate recommendations. Algorithm 1 summarizes our method.

Algorithm 1. *LwRec* Algorithm

Input: Ratings for n items by m users, values k and z

1 Initialize $\Theta_{m\times p}$, where $p = \frac{n!}{(n-k)!}$ (See Equation 3)
2 Randomly initialize $U_{z\times m}, G_{z\times p}, \alpha_m$ and β_m with non-negative values
3 **repeat**
4 $\quad|\quad$ Update U, G, α and β according to Equation 9
5 **until** *Reach convergence or the max iteration;*
6 **return** $U^\top G$

---

5   Experimental Setup

5.1   Datasets

For our experiments, we use two MovieLens[1] data sets: MovieLens-100K and MovieLens-1M. MovieLens-100K dataset contains 100,000 ratings given by 943

---

[1] http://grouplens.org/datasets/movielens/.

users on 1682 movies. MovieLens-1M dataset is larger with 1,000,000 ratings given by 6040 users on 3952 movies. In MovieLens-100K as well as MovieLens-1M the ratings are given on an integer scale from 1 to 5. For both the datasets we assign 10 ratings for each user for testing and the rest for training.

5.2 *LwRec* Setup

For both datasets, we consider top 1 item rankings (i.e. $k = 1$), since the topmost position in a ranking is the most important one. Moreover, it also makes our experiments computationally inexpensive since when $k = 1$, $p = \frac{n!}{(n-1)!} = n$ (number of items). A higher value of k, would make the value of p huge. For example, for MovieLens-1M ($n = 3952$), when $k = 2$, $p \approx 1.56 \times 10^7$ and for $k = 3$, $p \approx 6.17 \times 10^{10}$. Probably higher values of k could result in some performance gains, but in this study we restrict the scope to experiment with $k = 1$. Moreover, for the matrices U and G, we have used the column length of 10 (i.e. $z = 10$).

We generate the probability distributions of known item rankings i.e. Θ using the training set and then generate the matrix of predicted probabilities i.e. $U^{\top}G$ (Algorithm 1). Since we use $k = 1$, each row of $U^{\top}G$ would contain probabilities for n items (as each item ranking has only one item in this case). Therefore, items in the test set can be simply ordered by their decreasing predicted probabilities.

5.3 Baselines

We used the following state-of-the-art algorithms as our comparison partners:
1. CF: CF [2] calculates the similarity between users, and ranks the items according to the predicted ratings for each user.
2. Matrix Factorization (MF): User matrix U and item matrix I are optimized in MF [10], to minimize the difference between their product UI^T and rating matrix R. UI^T regenerates the rating matrix to predict unknown ratings.
3. EigenRank: EigenRank [15] is a pair-wise ranking-oriented algorithm that employs a greedy aggregation method to aggregate the predicted pairwise preferences of items into total ranking.
4. ListRankMF: ListRankMF [22] minimizes a loss function representing uncertainty between training and output lists produced by a MF ranking model.
5. FISM: Factored Item Similarity Models (FISM) [9] learn the item-item similarity matrix as a product of two low-dimensional latent factor matrices. While **FISMrmse** computes loss using squared error loss function, **FISMauc** considers a ranking error based loss function.
6. LLORMA: Local Low-Rank Matrix Approximation (LLORMA) [13] approximates the observed matrix as a weighted sum.
7. ListCF: ListCF [7], a ranking oriented CF algorithm, predicts item order for a user, based on similar users probability distributions over item permutations.

5.4   Evaluation Metrics

We use the standard ranking accuracy metric called normalized discounted cumulative gain (NDCG@1-10) [8] that is able to handle multiple levels of relevance, to evaluate item rankings generated by *LwRec* and the baselines.

Statistical significance of observed differences between the performance of two runs is tested using a two-tailed paired t-test and is denoted using ▲ (or ▼) for strong significance for $\alpha = 0.01$; or ᐃ (or ᐁ) for weak significance for $\alpha = 0.05$.

6   Results

In Table 1, we can see that *LwRec* outperforms the comparison partners for all the metrics (NDCG@1 to 10) for MovieLens-100K as well as MovieLens-1M. ListCF is the second best followed by LLORMA and FISMrmse, for both datasets. For MovieLens-100K, EigenRank and ListRankMF have comparable performances followed by MF and FISMauc. For MovieLens-1M, ListRankMF performs better than FISMauc followed by MF.

Table 1. Ranking performance of *LwRec* against baselines

Statistical significance shown for *LwRec* against LLORMA

Performance for MovieLens-100K									
NDCG	@1	@2	@3	@4	@5	@6	@7	@8	@9	@10
CF	0.5990	0.6394	0.6707	0.6938	0.7182	0.7442	0.7705	0.7970	0.8245	0.8546
MF	0.6629	0.6711	0.6918	0.7158	0.7373	0.7651	0.7895	0.8154	0.8418	0.8683
EigenRank	0.6734	0.6799	0.6972	0.7192	0.7408	0.7634	0.7889	0.8146	0.8407	0.8701
ListRankMF	0.6769	0.6792	0.6989	0.7140	0.7316	0.7532	0.7772	0.8057	0.8368	0.8684
FISMauc	0.6480	0.6681	0.6912	0.7132	0.7363	0.7598	0.7826	0.8086	0.8360	0.8661
FISMrmse	0.6735	0.6868	0.7060	0.7246	0.7475	0.7684	0.7914	0.8164	0.8431	0.8726
LLORMA	0.6794	0.6898	0.7092	0.7264	0.7488	0.7705	0.7950	0.8219	0.8462	0.8738
ListCF	0.6846	0.6897	0.7100	0.7274	0.7500	0.7732	0.7982	0.8243	0.8499	0.8752
LwRec	**0.6930**	**0.6991**	**0.7200**ᐃ	**0.7422**▲	**0.7643**▲	**0.7844**▲	**0.8059**▲	**0.8287**	**0.8527**	**0.8801**ᐃ

Performance for MovieLens-1M									
NDCG	@1	@2	@3	@4	@5	@6	@7	@8	@9	@10
CF	0.6214	0.6498	0.6710	0.6954	0.7189	0.7437	0.7708	0.7981	0.8272	0.8589
MF	0.6619	0.6649	0.6802	0.7008	0.7238	0.7483	0.7741	0.8026	0.8322	0.8642
EigenRank	0.6486	0.6571	0.6746	0.6958	0.7190	0.7428	0.7688	0.7966	0.8268	0.8608
ListRankMF	0.7084	0.7078	0.7203	0.7342	0.7532	0.7736	0.7972	0.8225	0.8498	0.8803
FISMauc	0.6784	0.6951	0.7109	0.7315	0.7526	0.7750	0.7983	0.8235	0.8493	0.8770
FISMrmse	0.7157	0.7178	0.7279	0.7440	0.7634	0.7849	0.8071	0.8315	0.8569	0.8847
LLORMA	0.7116	0.7174	0.7303	0.7479	0.7672	0.7878	0.8100	0.8340	0.8587	0.8854
ListCF	0.7204	0.7243	0.7359	0.7504	0.7685	0.7895	0.8136	0.8384	0.8627	0.8876
LwRec	**0.7204**	**0.7281**	**0.7428**▲	**0.7600**▲	**0.7777**▲	**0.7988**▲	**0.8207**▲	**0.8436**▲	**0.8667**▲	**0.8906**▲

We also calculate statistical significance of *LwRec* against ListCF which is our best performing comparison algorithm. The results for MovieLens-100K show weak to strong statistical significance for most metrics and for MovieLens-1M the results have strong statistical significance in almost all cases.

7   Conclusion

In this paper, we defined a novel listwise user-ranking probability prediction problem. Then we described *LwRec*, a listwise recommendation algorithm, that solves the problem by minimizing a listwise loss function using non-negative matrix factorization. Our experimental results on benchmark datasets show significant performance gains of *LwRec* over state-of-the-art recommender algorithms.

In this study, we have experimented for top k item rankings, for $k = 1$. In the future, we would like to explore the effect on results on using higher value of k. Moreover, we have used column length 10 for the matrices U and G. It would be interesting to see the changes in results on varying this column length.

References

1. Adomavicius, G., Tuzhilin, A.: Toward the next generation of recommender systems: a survey of the state-of-the-art and possible extensions. IEEE Trans. Knowl. Data Eng. **17**(6), 734–749 (2005)
2. Breese, J.S., Heckerman, D., Kadie, C.: Empirical analysis of predictive algorithms for collaborative filtering. In: UAI, pp. 43–52 (1998)
3. Cao, Z., Qin, T., Liu, T.Y., Tsai, M.F., Li, H.: Learning to rank: from pairwise approach to listwise approach. In: ICML, pp. 129–136 (2007)
4. Gopalan, P., Hofman, J.M., Blei, D.M.: Scalable recommendation with hierarchical poisson factorization. In: UAI, pp. 326–335 (2015)
5. Gunawardana, A., Shani, G.: A survey of accuracy evaluation metrics of recommendation tasks. J. Mach. Learn. Res. **10**, 2935–2962 (2009)
6. Herlocker, J., Konstan, J.A., Riedl, J.: An empirical analysis of design choices in neighborhood-based collaborative filtering algorithms. Inf. Retr. **5**(4), 287–310 (2002)
7. Huang, S., Wang, S., Liu, T.Y., Ma, J., Chen, Z., Veijalainen, J.: Listwise collaborative filtering. In: SIGIR, pp. 343–352 (2015)
8. Järvelin, K., Kekäläinen, J.: Cumulated gain-based evaluation of IR techniques. ACM Trans. Inf. Syst. **20**(4), 422–446 (2002)
9. Kabbur, S., Ning, X., Karypis, G.: FISM: Factored item similarity models for top-N recommender systems. In: SIGKDD, pp. 659–667 (2013)
10. Koren, Y., Bell, R., Volinsky, C.: Matrix factorization techniques for recommender systems. Computer **42**(8), 30–37 (2009)
11. Kullback, S.: Information Theory And Statistics. Dover Pubns, New York (1997)
12. Lee, D.D., Seung, H.S.: Algorithms for non-negative matrix factorization. In: NIPS, pp. 556–562 (2000)
13. Lee, J., Kim, S., Lebanon, G., Singer, Y.: Local low-rank matrix approximation. In: ICML, pp. II–82–II–90 (2013)
14. Linden, G., Smith, B., York, J.: Amazon.com recommendations: item-to-item collaborative filtering. IEEE Internet Comput. **7**(1), 76–80 (2003)

15. Liu, N.N., Yang, Q.: EigenRank: a ranking-oriented approach to collaborative filtering. In: SIGIR, pp. 83–90 (2008)
16. Marden, J.I.: Analyzing and Modeling Rank Data. CRC Press, London (1996)
17. Ning, X., Karypis, G.: SLIM: sparse linear methods for top-N recommender systems. In: ICDM, pp. 497–506 (2011)
18. Pan, W., Chen, L.: GBPR: group preference based Bayesian personalized ranking for one-class collaborative filtering. In: IJCAI, pp. 2691–2697 (2013)
19. Rendle, S.: Factorization Machines with libFM. ACM Trans. Intell. Syst. Technol. **3**(3), 57:1–57:22 (2012)
20. Salakhutdinov, R., Mnih, A.: Probabilistic matrix factorization. In: NIPS, pp. 1257–1264 (2007)
21. Shi, Y., Karatzoglou, A., Baltrunas, L., Larson, M., Oliver, N., Hanjalic, A.: CLiMF: learning to maximize reciprocal rank with collaborative less-is-more filtering. In: RecSys, pp. 139–146 (2012)
22. Shi, Y., Larson, M., Hanjalic, A.: List-wise learning to rank with matrix factorization for collaborative filtering. In: RecSys, pp. 269–272 (2010)
23. Wang, S., Sun, J., Gao, B.J., Ma, J.: VSRank: a novel framework for ranking-based collaborative filtering. ACM Trans. Intell. Syst. Technol. **5**(3), 51:1–51:24 (2014)
24. Weimer, M., Karatzoglou, A., Le, Q.V., Smola, A.: CofiRank: maximum margin matrix factorization for collaborative ranking. In: NIPS, pp. 1593–1600 (2007)

Estimation of Business Demography Statistics: A Method for Analyzing Job Creation and Destruction

Masao Takahashi[1]([⊠]) [iD], Mika Sato-Ilic[2] [iD], and Motoi Okamoto[3] [iD]

[1] Statistics Bureau, Ministry of Internal Affairs and Communications,
Shinjuku, Tokyo 162-8668, Japan
mtakahashi3@soumu.go.jp

[2] Faculty of Engineering, Information and Systems, University of Tsukuba,
Tsukuba, Ibaraki 305-8573, Japan
mika@risk.tsukuba.ac.jp

[3] The Institute of Statistical Mathematics,
Research Organization of Information and Systems,
Tachikawa, Tokyo 190-8562, Japan
mokamoto@ism.ac.jp

Abstract. Business demography statistics, which provide data on the numbers of births, deaths and survivals of enterprises and/or establishments in a specific period, serve as important information for policymakers who intend to make decisions on the policy to promote entrepreneurship, which is considered as an essential instrument for improving competitiveness and generating economic growth and job opportunities. These important statistics can be compiled from statistical business registers in many countries. In Japan, the statistical business register is being re-engineered not only by redesigning the system but also by improving the quality of data sources; however, it will take some years before business demography statistics can be compiled directly from the Japanese statistical business register. In this paper, an alternative method of estimating business demography statistics that can be obtained directly from the data of Economic Censuses is proposed. The model used here is basically based on previous works, however, an enhanced version of the model is proposed here so that quantitative attributes of establishments/enterprises such as the number of persons engaged can be analyzed. Based on the enhanced model, a numerical example on job creation and destruction is given using the micro data of Economic Censuses, which reveals the importance of fostering entrepreneurship as well as opening establishments as policy measures for economic growth and job opportunities.

Keywords: Business register · Business demography · Economic census

M. Takahashi—The views and opinions expressed in this paper are those of the authors and do not necessarily reflect the policies of the organization to which the authors belong or belonged. This work was supported by JSPS KAKENHI Grant Number JP16H02013.

© Springer International Publishing AG, part of Springer Nature 2019
I. Czarnowski et al. (Eds.): KES-IDT 2018, SIST 97, pp. 33–43, 2019.
https://doi.org/10.1007/978-3-319-92028-3_4

1 Introduction

Economic statistics play a key role in reasonable decision making by citizens as well as by the government. In this connection, statistical business registers (SBRs) have served as backbones in producing economic statistics in many countries for the past few decades. An SBR is a regularly updated, structured database of economic units, e.g. establishments and enterprises, in a territorial area, maintained by a National Statistical Institute (NSI), and used for statistical purposes [1]. There are a number of roles in an SBR, however, major roles include providing a sampling frame for various economic surveys and producing economic statistics directly from the SBR itself.

Among the economic statistics directly produced from an SBR, business demography statistics constitute a core element together with the basic statistics such as the number of units in the SBR. Business demography statistics provide data on the numbers of births and deaths of enterprises and/or establishments in a specific period, and on the number of enterprises or establishments that were born in a previous period and continued in (i.e. survived to) the specific period [2]. In this context business demography statistics serve as important information for policymakers who intend to make decisions on the policy to promote entrepreneurship which is considered as an essential instrument for improving competitiveness and generating economic growth and job opportunities [3].

In Japan, a statistical business register called "Establishment Frame Database (EFD)" has been operated since 2013 based on the Statistics Act. The major data source for the EFD is the Economic Census, which has been conducted twice every five years, while data from sample surveys and some administrative data such as the Labor Insurance Data, the Commercial and Corporate Registration Data, etc. are also used for updating the EFD in-between economic censuses.

In order to produce accurate and timely business demography statistics from the EFD, it is necessary that the data of the EFD be kept up-to-date as far as possible so that the EFD accurately reflects the status of the population. The administrative data used for updating the EFD, which constitute the major data source in-between economic censuses in terms of births and deaths of establishments/enterprises, cover a large part of the population of the EFD; however, some individual proprietorship businesses such as those with no employees are covered by neither source of the administrative data. In addition, there might be a certain amount of time lag between opening or closing of businesses and their registrations. Moreover, it often happens that closed businesses are not registered for a long time, even forever. These situations could cause some problems in the accuracy of the business demography statistics [4], if the statistics were produced from the EFD directly.

To cope with the above situation, the Statistics Bureau of Japan (SBJ) is starting redesign of the system and two initiatives for improving the quality of the statistical business register EFD. One is to improve the method of the Economic Census and the other is to introduce profiling for large enterprises. The SBJ plans to change the method of a part of the Economic Census into a rolling census, where openings and closings of businesses are surveyed over a couple of years to cover the whole of Japan, while the ordinary Economic Census will be conducted every five years. In addition, by profiling large enterprises, the SBJ will obtain basic information on large enterprise groups/enterprises, e.g. activities, turnover, number of employees, etc. of affiliated enterprises and/or establishments. These initiatives are expected to improve the quality of the EFD from the aspect not only of providing the population frame of establishments and enterprises but also of producing business demography statistics using the EFD. However, there are still some issues to be dealt with: basic information on the survived establishments detected in the course of the rolling census, such as activities and numbers of employees, should be estimated in some way because such establishments will only have their existence confirmed during the rolling census.

In this paper, we propose an alternative method of estimating business demography statistics from the perspective of risk engineering. We considered the elements of business demography statistics e.g. the death or transfer-out of establishments as the "risks" of the establishments and estimated the elements of the statistics by using not the data of the EFD but Economic Census data directly in order to cope with the above situation of the EFD. The method of this estimation is based on Takahashi [5] and Takahashi and Takabe [4], but we extended the model used there to be able to estimate job creation and destruction. In other words, we enhanced the method to estimate the annual increase rate of the number of persons engaged in newly opened and survived establishments by category such as industry, which have not been taken into account in the previous works. We define six variables concerning business demography statistics of establishments: annual opening rate, annual closing rate, annual surviving rate, annual activity transfer-in rate, annual activity transfer-out rate, and annual increase rate of the number of persons engaged in the surviving establishments. Then, by using the seven kinds of constant values derived from the micro data of two consecutive Economic Censuses, we have formulated a model comprising a system of equations. We finally obtained the method of estimating business demography statistics by solving the system of equations, by which we could estimate the abovementioned business demography statistics including annual rates of job creation and destruction.

The remainder of the paper is structured as follows. We begin by introducing related work on estimating business demography statistics by using the data of censuses and/or statistical surveys in Sect. 2. In Sect. 3, we introduce how to formulate our model for estimation. We also show the solution of the model formulae in the form of the system of equation to obtain the values of variables on business demography statistics. In Sect. 4, we describe a numerical example by using the real micro data of Economic Censuses. The last Section includes the conclusion and future work.

2 Related Work

There are a few kinds of works that tried to produce business demography statistics using the data of economic census and/or statistical surveys directly, though many countries, especially those in Europe, produce the statistics from their business registers or do not conduct an economic census.

Among the countries that conduct an economic census, Mexico produces business demography statistics from the data of their economic census. They have published the "Life expectancy of businesses in Mexico 2015" as part of business demography statistics, in which they performed longitudinal analysis that involves tracking the historical life lines of businesses captured from the 1989 up to the 2009 Economic Census (five censuses). They studied the demographic phenomena of survival and death as well as the life expectancy of businesses by area, business size, selected economic sector, etc. using the life table methodology [6], however, the study was only for the survival and death of businesses and did not include an analysis of quantitative attributes of businesses such as the number of persons engaged. In addition, the study did not take into account the transition between statuses such as the kinds of economic activity statuses, i.e. industry.

There are two kinds of studies on business demography based on economic censuses or statistical surveys in Japanese official statistics that analyze not only the births and deaths of businesses but also the quantitative attributes of businesses, such as the number of employees or the number of persons engaged, as part of business demography statistics.

Since 2011, the Ministry of Health, Labour and Welfare has been releasing the Indicators of Job Creation and Destruction based on the Survey on Employment Trends as a kind of business demography statistics. This Survey is a sample survey with about 15,000 sample establishments and is conducted twice every year. But care should be taken in using these Indicators because the scope of the Survey is limited to those establishments that employ 5 or more employees.

The Small and Medium Enterprise Agency in the Japanese Government has also produced a kind of business demography statistics in the 2011 White Paper on Small and Medium Enterprises in Japan, in which the rates of births and deaths of establishments by industry as well as the rates of job creation and destruction by industry between two consecutive Economic Census were studied. The calculation of the rates in this study, however, were performed under a rather simple assumption that the same number of establishments were opened/closed each year between the two consecutive Economic Censuses, while we suggest that annual opening/closing rates of establishments are the same between the two consecutive Economic Censuses, which reflects the reality to a larger extent.

3   Model Formulation

The model and the method for estimating business demography statistics used in this paper are based on those given in Takahashi [5] and Takahashi and Takabe [4]. But we extended the model so that business demography statistics on quantitative attributes of businesses such as the number of persons engaged in the business can be estimated by using the model. This enabled us to estimate job creation and destruction by industry, which constitute a key part of business demography statistics.

In the method, the data source for estimating business demography statistics is the data of the Economic Censuses, and the statistical unit for the estimation is an establishment, which is the basic enumeration unit in conducting the Economic Census in Japan. An establishment in the Japanese Economic Census corresponds to a local unit.

This method has an advantage that closed businesses can be certainly detected by the implementation of the Economic Census in which the list of businesses in the previous Census is utilized in the ongoing Census allowing for the detection of closed businesses. However, as pointed out in Ahmad [7], it is difficult to get information on short-lived births – businesses that opened after the previous Census and closed before the ongoing Census. This weakness can be minimized by relatively frequent implementation of the Economic Census in Japan where the Census has been conducted twice every five years.

3.1   Estimation Model

In estimating business demography statistics, we have selected six unknown variables in regard to births, deaths, survivals and transfers among categories which are related to the analyses concerned. Then, we set up six model formulae to form simultaneous equations using the above six unknown variables and known quantities obtained from two consecutive Economic Censuses. By solving the equations, the values of the six unknown variables were derived.

First of all, we defined six unknown variables as the indicators for business demography statistics of establishments, all of which are annual rates by the category concerned such as industry.

$$
\left.
\begin{array}{l}
R_b : \text{ increase rate of establishments/jobs by births of establishments} \\
R_d : \text{ decrease rate of establishments/jobs by deaths of establishments} \\
R_s : \text{ survival rate of establishments/jobs (in the same category)} \\
R_{ti} : \text{ transfer - in rate of establishments/jobs (of survivals)} \\
R_{to} : \text{ transfer - out rate of establishments/jobs (of survivals)} \\
R_{si} : \text{ ratio of the number of jobs of the survivals at the end of the term} \\
\qquad \text{ to that in one year before}
\end{array}
\right\} \quad (1)
$$

Here each rate/ratio is applicable both for the number of establishments and the quantitative attributes such as jobs. In addition, the above "jobs" can be replaced by other quantitative attributes such as turnover, costs and value added. It should be noted when the numbers of establishments are the values of concern, R_{si}, which we have introduced in this study, should be set to one, because R_{si} denotes the ratio of the number of an attribute such as jobs of the survived establishments at the end of the term to the number of the attribute one year before.

In order to obtain the value of the above six unknown variables (1), we set up six model formulae using the following seven known quantities derived from the micro data of the Economic Censuses. Note that the following values are considered by each category such as industry.

N_p : number of establishments/jobs in the previous Economic Census
N_l : number of establishments/jobs in the last Economic Census
N_d : number of establishments/jobs decreased by the death (closing) of establishments between

    the previous and the last Economic Census

N_{tsl} : number of surviving establishments/jobs at the last Economic Census including

    the transferred - in establishments to the category

N_{sp} : number of surviving establishments/jobs in the same category in terms of

    the previous Economic Census

N_{sl} : number of surviving establishments/jobs in the same category in terms of the last Economic Census
n : number of years between the previous and the last Economic Census

$$\left. \vphantom{\begin{array}{c} 1 \\ 2 \\ 3 \\ 4 \\ 5 \\ 6 \\ 7 \\ 8 \end{array}} \right\} \quad (2)$$

Based on the various relationships among the six unknown variables (1) and the seven known quantities (2), we formulated, in this study, six model formulae as follows.

The first model formula can be obtained from the relation of the number of establishments/jobs between the previous and the last Economic Census. The number of establishments/jobs at the last Census N_l ca be estimated by that at the previous census N_p multiplied by the n th power of annual increase rate calculated by $(R_s \cdot R_{si} + R_{ti} \cdot R_{si} + R_b)$, which is as follows.

$$N_p \cdot (R_s \cdot R_{si} + R_{ti} \cdot R_{si} + R_b)^n = N_l \qquad (3)$$

The second formula relates to the survivals of establishments/jobs between two Economic Censuses. As the number of surviving establishments/jobs in terms of the previous Economic Census, N_{sp}, can be calculated by the number of establishments/jobs at the previous Census, N_p, multiplied by the n th power of the survival rate R_s,

$$N_p \cdot R_s^n = N_{sp}. \qquad (4)$$

The third formula is another model concerning the survivals of establishments/jobs between two Economic Censuses; the number of establishments/jobs at the previous Census, N_p, multiplied by the n th power of the survival rate in consideration of its annual increase/decrease results in the number of surviving establishments/jobs in terms of the last Economic Census, N_{sl}.

$$N_p \cdot (R_s \cdot R_{si})^n = N_{sl} \qquad (5)$$

Here, we temporarily define the number of surviving establishments/jobs transferred into the category concerned as N_{ti}, the following equation holds, both sides of which denote the number of survived establishments/jobs including those transferred-in.

$$N_{sl} + N_{ti} = N_{tsl} \qquad (6)$$

Here, N_{ti} can be estimated by summing up the number of establishments/jobs transferred into the category concerned every year in consideration of the survival R_s and increase/decrease of quantitative attributes R_{si}, which is shown below.

$$
\begin{aligned}
N_{ti} &= N_p \cdot R_{ti} \cdot R_{si} \cdot (R_s \cdot R_{si})^{(n-1)} + N_p \cdot (R_s \cdot R_{si} + R_{ti} \cdot R_{si} + R_b) \cdot R_{ti} \cdot R_{si} \cdot \\
&\quad (R_s \cdot R_{si})^{(n-2)} + N_p \cdot (R_s \cdot R_{si} + R_{ti} \cdot R_{si} + R_b)^2 R_{ti} \cdot R_{si} \cdot (R_s \cdot R_{si})^{(n-3)} + \cdots + \\
&\quad N_p \cdot (R_s \cdot R_{si} + R_{ti} \cdot R_{si} + R_b)^{n-1} \cdot R_{ti} \cdot R_{si} \\
&= N_p \cdot \left\{ \frac{(R_s \cdot R_{si})^n - (R_s \cdot R_{si} + R_{ti} \cdot R_{si} + R_b)^n}{R_s \cdot R_{si} - (R_s \cdot R_{si} + R_{ti} \cdot R_{si} + R_b)} \right\} \cdot R_{ti} \cdot R_{si}
\end{aligned}
\tag{7}
$$

So the fourth formula (6) is written as

$$
N_{sl} + N_p \cdot \left\{ \frac{(R_s \cdot R_{si})^n - (R_s \cdot R_{si} + R_{ti} \cdot R_{si} + R_b)^n}{R_s \cdot R_{si} - (R_s \cdot R_{si} + R_{ti} \cdot R_{si} + R_b)} \right\} \cdot R_{ti} \cdot R_{si} = N_{tsl}.
\tag{8}
$$

The next formula, obtained by summing up the number of dead (closed) establishments/lost jobs every year, is

$$
N_p \cdot R_d + N_p \cdot R_s \cdot R_d + N_p \cdot R_s^2 \cdot R_d + \cdots + N_p \cdot R_s^{n-1} \cdot R_d = N_d,
\tag{9}
$$

which leads to the fifth formula,

$$
N_p \cdot \left(\frac{1 - R_s^n}{1 - R_s} \right) \cdot R_d = N_d.
\tag{10}
$$

The last formula is based on the "conservation of the number of establishments/jobs." As every establishment in a category transits to a status of survived, dead (closed), or transferred-out one year later, the following equation holds.

$$
N_p = N_p \cdot R_s + N_p \cdot R_d + N_p \cdot R_{to}
\tag{11}
$$

Dividing both sides of the above formula by N_p and replacing both sides, the Eq. (11) can be simplified to

$$
R_s + R_d + R_{to} = 1.
\tag{12}
$$

In summary, we have derived six model formulae for estimating business demography statistics as equations of (3), (4), (5), (8), (10) and (12).

3.2   The Solution

By solving the system of six Eqs. (3), (4), (5), (8), (10) and (12) that form the model formulae, the following values for calculating business demography statistics are obtained for the six unknown variables.

$$
\left.\begin{aligned}
R_s &= \left(N_{sp}/N_p\right)^{1/n} \\
R_{si} &= \left(N_{sl}/N_{sp}\right)^{1/n} \\
R_{ti} &= \left(\frac{N_{tsl} - N_{sl}}{N_p}\right) \cdot \left\{\frac{\left(N_{sl}/N_p\right)^{1/n} - \left(N_l/N_p\right)^{1/n}}{N_{sl}/N_p - N_l/N_p}\right\} \cdot \left(\frac{N_{sp}}{N_{sl}}\right)^{1/n} \\
R_b &= \left(N_l/N_p\right)^{1/n} - R_s \cdot R_{si} - R_{ti} \cdot R_{si} \\
R_d &= \frac{N_d}{N_p} \cdot \frac{1 - R_s}{1 - R_s^n} \\
R_{to} &= 1 - R_s - R_d
\end{aligned}\right\} \tag{13}
$$

In performing the above calculations to compile business demography statistics, the results of the consecutive two Economic Censuses were used. The format for the calculation is shown in Table 1. In this table, six of the known quantities N_p, N_l, N_d, N_{tsl}, N_{sp}, and N_{sl} for the category A are shown. The remaining quantity n can be derived from the census dates of the two consecutive Economic Censuses.

Table 1. Relation table for compiling business demography statistics

(The number of years between the previous and the last Economic Census: n)

Last Census / Last − 1 (Previous) Census		Surviving establishments					Closed (dead) establishments	Total
		Category A	B	.....	R	Sub-total		
Surviving establishments	Category A	N_{sl} N_{sp}					N_d	N_p
	B							
	⋮							
	R							
	Sub-total	N_{tsl}						
Opened(born) establishments								
Total		N_l						

4   A Numerical Example

Based on the above model formulae, we estimated business demography statistics concerning job creation and destruction by using the data of two consecutive Economic Censuses conducted in July 1, 2009 and February 1, 2012. We produced the table of the number of persons engaged in businesses as indicated in Table 1, and derived seven

Table 2. Indicators of business demography statistics by industry (major category) – average annual rates between July 1, 2009 and February 1, 2012

Industrial classification	R_s	R_b	R_{ti}	$R_{si}-1$	$R_d{}^*$	$R_{to}{}^*$	(R_i)
Total	0.936	0.018	0.017	0.012	−0.048	−0.016	−0.018
A: Agriculture and forestry	0.930	0.021	0.021	0.010	−0.045	−0.026	−0.019
B: Fisheries	0.933	0.013	0.018	−0.002	−0.053	−0.014	−0.038
C: Mining and quarrying of stone and gravel	0.811	0.005	0.077	−0.025	−0.041	−0.148	−0.133
D: Construction	0.944	0.008	0.016	−0.010	−0.041	−0.015	−0.041
E: Manufacturing	0.937	0.008	0.019	0.013	−0.038	−0.025	−0.023
F: Electricity, gas, heat supply and water	0.979	0.010	0.003	−0.005	−0.012	−0.009	−0.013
G: Information and communications	0.907	0.016	0.027	0.030	−0.068	−0.025	−0.021
H: Transport and postal activities	0.956	0.010	0.011	−0.007	−0.037	−0.007	−0.030
I: Wholesale and retail trade	0.926	0.023	0.019	0.002	−0.054	−0.020	−0.030
J: Finance and insurance	0.942	0.024	0.015	0.017	−0.055	−0.003	−0.002
K: Real estate and goods rental and leasing	0.923	0.012	0.031	0.015	−0.054	−0.023	−0.018
L: Scientific research, professional and technical services	0.906	0.016	0.035	0.018	−0.059	−0.036	−0.025
M: Accommodations, eating and drinking services	0.918	0.037	0.009	0.017	−0.075	−0.007	−0.019
N: Living-related and personal services and amusement services	0.936	0.020	0.013	0.007	−0.051	−0.012	−0.024
O: Education, learning support	0.957	0.016	0.006	0.020	−0.036	−0.007	−0.001
P: Medical, health care and welfare	0.973	0.023	0.002	0.039	−0.025	−0.002	−0.037
Q: Compound services	0.922	0.003	0.011	0.000	−0.014	−0.064	−0.064
R: Services, n.e.c.	0.926	0.015	0.036	0.018	−0.055	−0.019	−0.005

* The signs of R_d and R_{to} are reversed for convenience.

constant values for each major category of industrial classification for the Economic Census, which is compatible with the Japan Standard Industry Classification. By assigning the seven values to the formulae of (13), we obtained the values of six values of business demography statistics together with the annual average increase rate of jobs (R_i), which are shown in Table 2.

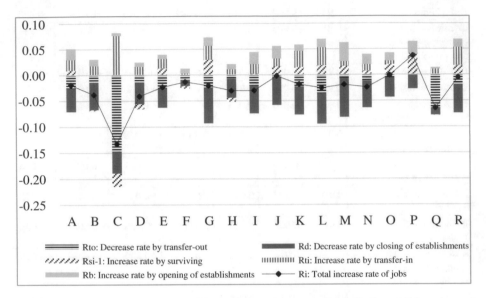

Fig. 1. Increase rate of jobs by opening, by transfer-in, by surviving of establishments, and decrease rate of jobs by closing, by transfer-out of establishments

Figure 1 is an illustration of the indicators of business demography statistics of job creations, destructions, survivals, etc. that are shown in Table 2. As a result, we found in a quantitative way that the annual job creation rate by opening of businesses, R_b exceeds that in surviving establishments ($R_{si}-1$) in many industries. This implies the importance of fostering entrepreneurship as well as opening establishments as policy measures for economic growth and job opportunities.

5   Conclusions

We have proposed the method for estimating business demography statistics, directly from the data of economic censuses, by category such as industry, and by factor such as opening, closing and surviving of establishments in terms of annual increase/decrease rates. Based on the proposed model formulae, we performed a calculation of job creation and destruction, which constitute the indicators of business demography statistics, using the micro data of two consecutive Economic Censuses conducted in 2009 and 2012.

The results we derived by the calculation are expected to contribute to various analyses in this field. Moreover, the results can be utilized as a kind of benchmark when authentic business demography statistics will be produced from the statistical business register in the future.

References

1. United Nations Economic Commission for Europe (UNECE): Guidelines on Statistical Business Registers. ECE/CES/39, New York and Geneva, p. 1 (2015)
2. United Nations Economic Commission for Europe (UNECE): Guidelines on Statistical Business Registers. ECE/CES/39, New York and Geneva, p. 100 (2015)
3. Eurostat and OECD: Eurostat-OECD Manual on Business Demography Statistics, p. 9 (2007)
4. Takahashi, M., Takabe, I.: Application of an alternative method for compiling business demography statistics of establishments. In: Proceedings of the 60th ISI World Statistics Congress, International Statistical Institute, Rio de Janeiro, pp. 1568–1573 (2015)
5. Takahashi, M.: Provisional calculation of the rates of births and deaths of establishments by industry based on the results of the Establishments and Enterprise Census. In: Report of the Meeting of Japan Statistical Society, pp. 67–68. Japan Statistical Society, Tokyo (2005). (in Japanese)
6. Blancas, A.: Business demography in mexico. progress and perspectives. In: Meeting of the Group of Experts on Business Registers. United Nations Economic Commission for Europe, Brussels (2015). http://www.unece.org/fileadmin/DAM/stats/documents/ece/ces/ge.42/2015/Session_I_M%C3%A9xico_-_Business_Demography.pdf. Accessed 2 Dec 2017
7. Ahmad, N.: A proposed framework for business demography statistics. OECD Statistics Working Papers, 2006/03. OECD Publishing, Paris (2006)

Multivariate Multiple Orthogonal Linear Regression

Thi Binh An Duong[1(✉)], Jun Tsuchida[1], and Hiroshi Yadohisa[2]

[1] Graduate School of Culture and Information Science,
Doshisha University, Kyoto, Japan
eiq1004@mail4.doshisha.ac.jp
[2] Department of Culture and Information Science,
Doshisha University, Kyoto, Japan

Abstract. We propose a multivariate multiple orthogonal linear regression (MMOLR) model. The MMOLR model expresses the relationship between two sets of dependent variables and independent variables. It is possible to use the MMOLR as a step to be followed by multivariate linear regression to compare and investigate the relationships between dependent variables, which are still limited in the multivariate linear regression model. The MMOLR takes into account the advantages of the errors-in-variables model, that is, total least squares, and thereby, examines the errors of all independent and dependent variables. Consequently, the assumptions of the model are easy to satisfy in practice. Moreover, in the context of total least squares, we reveal the relationship between the MMOLR and the canonical regression model.

Keywords: Canonical correlation analysis
Ordinary least squares · Total least squares

1 Introduction

Multivariate analysis is a set of models, or statistical techniques, which researchers widely use and apply in many fields [1]. Some common ones are used to analyze two set of variables, such as canonical regression, two-block partial least squares, and multivariate linear regression. The choice of method depends on the type of data and the information needs of the researchers [2].

Canonical regression, also known as canonical correlation analysis (CCA), is a model that expresses the relationship between two sets of independent variables (IVs) and dependent variables (DVs) [3–5]. The CCA is an extension of ordinary least square (OLS) regression [6] and has numerous applications, such as testing the omnibus null hypothesis, assessing overall model fit, testing composite hypotheses, and model validation [6]. The CCA focuses on correlation than on regression, and thus, the term "canonical correlation" is more commonly used.

© Springer International Publishing AG, part of Springer Nature 2019
I. Czarnowski et al. (Eds.): KES-IDT 2018, SIST 97, pp. 44–53, 2019.
https://doi.org/10.1007/978-3-319-92028-3_5

In contrast to the CCA, the two-block partial least squares method (2B-PLS) introduced by Wold (1975) (see [7]) is used to analyze the covariance between two sets of variables [8]. In other words, 2B-PLS describes the relationship between the latent variables and latent constructs of two variables. It is possible to handle the collinearity problem by extracting the latent variables when applying this method to linear regression [9] and is consistent with the large number of predictor variables. Moreover, notably, there is no satisfactory explanation of the relationship between 2B-PLS and other multiple-response regression methods thus far [10].

The aforementioned limitations explain why the interpretations of these two methods have many practical limitations. The more popular and less limited interpretation in application involves multivariate linear regression (MvLR), which is a prediction model in which the values of responses can be predicted from a set of predictors [11,12]. In addition, the model is used to estimate linear association between predictors and responses [13]. However, each individual DV in this model is regressed separately on the IVs. Therefore, the inter-correlation between DVs is not examined [14]. From the perspective of application, the variables generally measure related aspects, and thus, the relational view of the DVs in a discrete way would lack merit and reduce the significance of analyzing the variables in sets. Moreover, for considering the association between IDs and DVs, the interpretation of a large number of coefficients simultaneously is unwieldy [15].

All three models are in the OLS approach, which considers only errors in IVs. According to [16,17], it is difficult to hold this assumption strictly in practice. To extend the interpretation by the total least squares (TLS) approach, this study proposes a multivariate multiple orthogonal linear regression model (MMOLR) that can describe linear relationships between two sets of variables, such as MvLR, and the relationship between DVs that are integrated in the model. Because of its canonical coefficients, the proposed model is very similar to canonical regression. We reveal the relationship between MMOLR and canonical regression in the sense of TLS.

The remainder of the paper is organized as follows. In the next section, we present the proposed model and coefficient estimation. In the third section, we compare the MMOLR and canonical regression in detail. The fourth section contains a numerical example while we include a practical example in the fifth section. Finally, we present the limitations of the research and its conclusion.

2 Multivariate Multiple Orthogonal Linear Regression

In the MMOLR model, the word "multivariate" means that there are many DVs, multiple implies a large number of IVs, and orthogonal linear regression is the model's method.

The following equation describes the MMOLR model (intercept excluded):

$$(\boldsymbol{Y} - \boldsymbol{E}_y)\,\boldsymbol{\beta}_y = (\boldsymbol{X} - \boldsymbol{E}_x)\,\boldsymbol{\beta}_x \qquad (1)$$

where:

$$Y = (y_1, y_2, \ldots, y_q), \quad X = (x_1, x_2, \ldots, x_p),$$
$$E_y = (\varepsilon_{y_1}, \varepsilon_{y_2}, \cdots \varepsilon_{y_q}), \quad E_x = (\varepsilon_{x_1}, \varepsilon_{x_2}, \ldots \varepsilon_{x_p}),$$
$$\beta_y = (\beta_{y_1}, \beta_{y_2}, \ldots, \beta_{y_q})^T, \quad \beta_x = (\beta_{x_1}, \beta_{x_2}, \ldots, \beta_{x_q})^T,$$
$$x_i, y_j, \varepsilon_{x_i}, \varepsilon_{y_j} \in \mathbb{R}^m, \quad (i = 1, 2, \ldots, p; \ j = 1, 2, \ldots, q).$$

The components and relationships of the Eq. (1) are schematically depicted in Fig. 1.

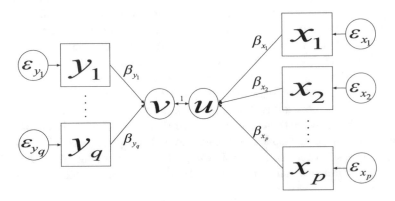

Fig. 1. Path diagram of multivariate multiple orthogonal linear regression

In order to estimate the coefficients, (1) is transformed by setting data matrix $A = (Y, X) = (a_1, a_2, \ldots, a_k) = (f_1, f_2, \ldots, f_m)^T$ with k column vectors a_j and m row vectors f_i; coefficient matrix $\beta = (\beta_y, -\beta_x)^T = (\beta_1, \beta_2, \ldots, \beta_k)^T$ and error matrix $E = (E_y, E_x) = (\varepsilon_1, \varepsilon_2, \ldots, \varepsilon_k)$ with $k = p + q$. Then, we rewrite the Eq. (1) as follows:

$$A\beta = E\beta. \tag{2}$$

In addition, we rewrite the Eq. (2) as

$$\beta_1 a_1 + \beta_2 a_2 + \cdots + \beta_k a_k = \varepsilon, \tag{3}$$

where $\varepsilon = \beta_1 \varepsilon_1 + \beta_2 \varepsilon_2 + \cdots + \beta_k \varepsilon_k$.

Equation (3) describes the general expression of the hyperplane H with normal vector $(\beta_1, \beta_2, \ldots, \beta_k)$. Thus, we can assume that the estimation of coefficients is to find H, such that errors from data points to perpendicular projections corresponding to H are the smallest. Furthermore, H can be written as $Pf_i = \alpha_i$, where P is a projection matrix with orthogonal rows, and hence, $PP^T = I$. Moreover, the square distance of f to H is $\| Pf_i \|^2$. Then, the objective function $g(P)$ is

$$g(P) = \sum_{i=1}^{m} \| Pf_i \|^2 \rightarrow minimize \tag{4}$$
$$\text{Subject to } PP^T = I$$

The solution of (4) is a matrix whose column vectors are eigenvectors corresponding to the smallest eigenvalues of the covariance matrix of $\boldsymbol{A}$ [18,19]. Moreover, [20] argue that singular value decomposition (SVD) can be used to solve the least square problem. [22] go further by claiming that the eigenvector corresponding to the smallest eigenvalue (denoted by $\boldsymbol{n}$) of SVD is the solution for the TLS problem. Therefore, orthogonal regression deals with the directions of $\boldsymbol{n}$ [19]. In other words, errors are minimized if projecting data to H with the direction of $\boldsymbol{n}$, that is, $\boldsymbol{n}$ is a normal vector of H. Thus, $\boldsymbol{n} = (n_1, n_2, \ldots, n_k) = (\beta_1, \beta_2 \ldots, \beta_k)$. However, it is noteworthy that $\boldsymbol{n}$ is very sensitive to outliers [21], and hence, when finding $\boldsymbol{n}$, it is better to use robust covariance matrixes.

3   Relationship Between the Proposed Model and the Canonical Regression Model

The dereference point between the objective function of the proposed model and canonical regression is optimizing design. Canonical regression maximizes the correlation or covariance between IVs and DVs. On the other hand, the proposed model minimizes it. This difference is caused by the following equation:

$$\text{tr}(\boldsymbol{\beta}^T (\boldsymbol{A} + \boldsymbol{E})^T (\boldsymbol{A} - \boldsymbol{E})\boldsymbol{\beta}) = \|\boldsymbol{A}\boldsymbol{\beta}\|^2 - \|\boldsymbol{E}\boldsymbol{\beta}\|^2. \tag{5}$$

From Eq. (5), we can derive the objective function for the maximizing Eq. (5) by two types of design. The first is by maximizing $\|\boldsymbol{A}\boldsymbol{\beta}\|^2$. The second is by minimizing $\|\boldsymbol{E}\boldsymbol{\beta}\|^2$. We extend the objective function of the proposed method to include canonical regression in the context of TLS. The objective function g extended from the proposed method is defined as follows:

$$g(\boldsymbol{\beta}) = \|\boldsymbol{E}\boldsymbol{\beta}\|^2 \to \text{minimize}$$
$$\text{Subject to } \boldsymbol{\beta}^T \boldsymbol{\beta} = 1, \ \boldsymbol{\beta}^T \boldsymbol{A}^T \boldsymbol{A} \boldsymbol{\beta} \geq r.$$

By using the Lagrange multiplier and assumption of the linear space spanned by $\boldsymbol{\beta}$ and $\boldsymbol{E}$ are orthogonal and direct sum, that is, $\boldsymbol{\beta} \dot{\oplus} \boldsymbol{E}$, the objective function g is rewritten as follows:

$$g(\boldsymbol{\beta}) = \|\boldsymbol{E}\boldsymbol{\beta}\|^2 - \lambda \|\boldsymbol{A}\boldsymbol{\beta}\|^2 \to \text{minimize}$$
$$\text{Subject to } \boldsymbol{\beta}^T \boldsymbol{\beta} = 1,$$

where $\lambda \geq 0$ represents tuning parameters. The role of tuning parameters λ is to adjust the maximizing $\|\boldsymbol{A}\boldsymbol{\beta}\|^2$ or minimizing $\|\boldsymbol{E}\boldsymbol{\beta}\|^2$. When $\lambda = 0$, the objective function g is the same as that of the proposed model. On the other hand, g is equivalent to canonical regression when we set a sufficiently larger λ.

From this fact, the objective function g is an extension from the objective function of canonical regression and the MMOLR.

We adopt the algorithm for alternative least squares to optimize g. Given $\boldsymbol{E}$, the estimator of $\boldsymbol{\beta}$ satisfies the following equation:

$$(\boldsymbol{E}^T \boldsymbol{E} - \lambda \boldsymbol{A}^T \boldsymbol{A})\boldsymbol{\beta} = \eta \boldsymbol{\beta}, \tag{6}$$

where η is the Lagrange multiplier. Equation (6) is the same as the eigenvalue problem. Therefore, we obtain the estimator by the eigenvector whose eigenvalue is the minimum. On the other hand, given $\boldsymbol{\beta}$, the estimator of $\boldsymbol{E}$ satisfies the following equation:

$$\boldsymbol{E}^T \boldsymbol{\beta}\boldsymbol{\beta}^T = \boldsymbol{O}. \tag{7}$$

Equation (7) indicates that the error matrix is orthogonal from $\boldsymbol{\beta}\boldsymbol{\beta}^T$, that is, the space of error is the orthogonal space of $\boldsymbol{\beta}$. We set $\boldsymbol{E} = \boldsymbol{A} - \boldsymbol{A}\boldsymbol{\beta}\boldsymbol{\beta}^T$, although it is arbitrary.

As mentioned above, we define the algorithm of the extended proposed model as follows:

Step 1. Set $\boldsymbol{E} = \boldsymbol{A}$.
Step 2. Update $\boldsymbol{\beta}$.
Step 3. Update $\boldsymbol{E}$.
Step 4. Repeat Steps 2 and 3 until $\boldsymbol{\beta}$ converge.

4   Numerical Example

In this section, we use the prediction error as a criterion to compare the proposed model with CCA and Canonical covariance. These two models can be considered as special cases of MMOLR. We evaluate the variance and bias of estimators through the prediction error.

The simulation data is generated in following steps. First, coefficient $\boldsymbol{\beta} = (\boldsymbol{\beta}_y^T, -\boldsymbol{\beta}_x^T)$ is generated in condition of $\beta_j \overset{i.i.d.}{\sim} U(-1,\ 1)$ $(i = 1,\ 2,\ \cdots,\ k)$, where $U(-1,\ 1)$ shows uniform distribution from -1 to 1. Then, IVs $\boldsymbol{x}_i \overset{i.i.d.}{\sim} N(\boldsymbol{0}_p,\ \sigma^2 \boldsymbol{I}_p)(i = 1,\ 2,\ \cdots,\ m)$, where $\boldsymbol{0}_p$ is p dimensional vector whose elements is 0, $\boldsymbol{I}_p$ is p dimensional identical matrix, and m is the number of objects. In this example, we set the number of observations at 300. Dependent variable $\boldsymbol{y}_i$ is generated as $\boldsymbol{y}_i = \boldsymbol{x}\boldsymbol{\beta}_x\boldsymbol{\beta}_y^T$. Therefore, the relationship between $\boldsymbol{x}_i$ and $\boldsymbol{y}_i$ is the ideal relationship for regression. We set σ as 1, 2, the number of IVs $p = 7$ and the number of DVs $q = 5$, that is, $k = 12$. The error matrix is high correlated case and the error variables are generated $\boldsymbol{\varepsilon}_{xi} \overset{i.i.d.}{\sim} N(\boldsymbol{0},\ \boldsymbol{\Sigma}_x)$ and $\boldsymbol{\varepsilon}_{yi} \overset{i.i.d.}{\sim} N(\boldsymbol{0},\ \boldsymbol{\Sigma}_y)$. The $(i,\ j)$th element of $\boldsymbol{\Sigma}_x$ and $\boldsymbol{\Sigma}_y$, which is represented σ_{ij}^x, σ_{ij}^y, respectively, is made as follows:

$$\sigma_{ij}^x = r^{|i-j|}, \sigma_{o\ell}^y = r^{|o-\ell|}$$
$$(i,\ j = 1,\ 2,\ \cdots,\ p;\ o,\ \ell = 1,\ 2,\ \cdots,\ q),$$

where r is the parameter for correlations and is set as 0.5, 0.7, 0.9. We obtain the data $\boldsymbol{X}^*, \boldsymbol{Y}^*$ by $\boldsymbol{X}^* = \boldsymbol{X} + \boldsymbol{E}_x$ and $\boldsymbol{Y}^* = \boldsymbol{Y} + \boldsymbol{E}_x$

The criterion of evaluation is the prediction error, which is defined as follows:

$$\|(\boldsymbol{Y}_{pr}, \boldsymbol{X}_{pr})\boldsymbol{\beta} - (\boldsymbol{Y}_{pr}^*, \boldsymbol{X}_{pr}^*)\hat{\boldsymbol{\beta}}\|,$$

where Y_{pr}, X_{pr}, Y_{pr}^*, and X_{pr}^* is the test data obtained in the same way as Y, X, Y^*, and X^*, respectively. $\hat{\beta}$ is the value of the estimator. We set iteration times at 100.

Figure 2 shows the boxplots of the prediction error. Boxplots are, from left, CCA, canonical covariance analysis, and proposed model. Our model has the best result in all cases. Therefore, our model is more effective than the CCA, when errors are higher correlated.

Table 1 shows the mean prediction error in all cases. Blanked values show the standard deviation of prediction error. The highlighted cells indicate the best result under each condition. From Table 1, our model demonstrates the best result under all cases. Moreover, the standard deviation of CCoVA tends to be increasing when the correlation between error variables is higher. On the other hand, the standard deviation of proposed method is not increasing.

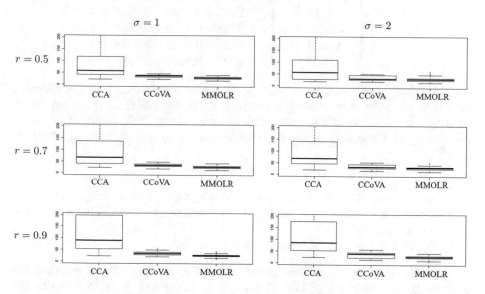

Fig. 2. Boxplots of square prediction errors

Table 1. The mean and standard deviation of prediction error

		CCA		CCoVA		MMOLR	
	r=0.5	176.355	(508.636)	35.440	(5.221)	29.053	(5.722)
$\sigma = 1$	r=0.7	215.599	(593.429)	34.566	(6.439)	26.818	(6.054)
	r=0.9	283.832	(743.962)	35.440	(6.927)	26.042	(5.050)
	r=0.5	565.195	(4629.857)	34.969	(9.731)	28.764	(8.861)
$\sigma = 2$	r=0.7	474.120	(2040.812)	35.046	(10.421)	28.728	(7.202)
	r=0.9	881.278	(5244.808)	34.578	(12.038)	25.897	(6.859)

5 Practical Example

From the perspective of application, we analyze a typical example of biochemical data extracted from the book "Multivariate Reduced-Rank Regression: Theory and Applications" (see [15] [Appendix 1]). The data include 33 observations with five DVs, comprising pigment creatinine (y_1), concentration of phosphate (y_2), phosphorus (y_3), creatinine (y_4), and choline (y_5), and three IVs, namely, weight (x_1), volume (x_2), and $x_3 = 100(specific\ gravity\text{-}1)$. In Chap. 1 of [15], the MvLR model was applied for these data to measure the involvement of IDs x_1, x_2, and x_3 in DVs y_1, y_2, y_3, y_4, and y_5.

The result of the MvLR model is summarized as follows:

$$\begin{pmatrix} y_1 \\ y_2 \\ y_3 \\ y_4 \\ y_5 \end{pmatrix} = \begin{pmatrix} 15.2809 \\ 1.4159 \\ 2.0187 \\ 1.8717 \\ -0.8902 \end{pmatrix} + \begin{pmatrix} -2.9090 & 1.9631 & 0.2043 \\ 0.6044 & -0.4816 & 0.2667 \\ 0.5768 & -0.4245 & -0.0401 \\ 0.6160 & -0.5781 & 0.3518 \\ 1.3798 & -0.6289 & 2.8908 \end{pmatrix} \begin{pmatrix} x_1 \\ x_2 \\ x_3 \end{pmatrix} \qquad (8)$$

According to the MvLR model, it is completely unknown how the variables of y interact with each other (all coefficients of the variables y are set to 1), although an analysis of the correlation matrix of the DVs shows them to be correlated, and the correlation between variables y_1, y_3, y_4 is especially high (see Fig. 3).

In addition, the following two questions arise.

- i. For an individual or the same individuals, that is, when the values of x_1, x_2, and x_3 are determined, what is the relationship between the variables of y_1, y_2, y_3, y_4, and y_5?
- ii. Which are the most affected or weakest DVs?

The MvLR model itself cannot compare DVs, although these variables are affected by the same set of IVs. Thus, it does not make sense to examine the relationships of variables in a set. We apply the MMOLR model to answer these questions. For comparing variables, the intercept is not considered, and the data are centered. In this example, the authors use a package that is robust to the "covRob" command in R to compute the covariance matrix of the data. The default of covRob allows auto select among Donoho–Stahel projection based estimators, the fast minimum covariance determinant algorithm of Rousseeuw and Van Driessen, and the orthogonalized quadrant correlation pairwise estimator for good estimates in a reasonable amount of time [23]. From the covariance matrix, eight eigenvalues and eigenvectors are computed. In addition, the smallest eigenvalue is 0.009 given eigenvector $(-0.004, -0.682, 0.634, -0.062, 0.005, 0.116, -0.008, 0.329)$. Thus, the equation expressing the relationship of the variables is as follows:

$$0.004y_1 + 0.682y_2 - 0.634y_3 + 0.062y_4 - 0.005y_5 = 0.116x_1 - 0.008x_2 + 0.329x_3 \quad (9)$$

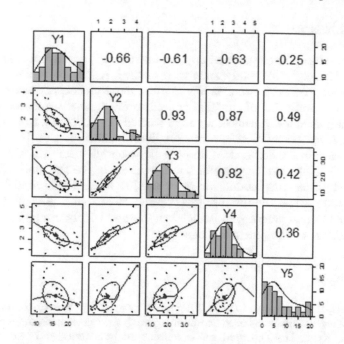

Fig. 3. Correlation of responses in biochemical data

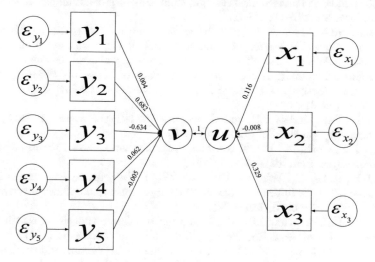

Fig. 4. Path diagram of the MMOLR model

The general result is depicted in Fig. 4.

With the same effect of the variables x_1, x_2, x_3, the DV y_2 is most sensitive to the changes of the set of IVs, while y_1 is the least sensitive variable.

6 Conclusion

The TLS has been developed and applied for decades, and is not a new method of fitting. Although this study aims to propose a regression model to describe the relationship between two sets of variables, more precisely, its contribution is to specify an application of the TLS as well as orthogonal regression.

The analysis model in this study is suggested as a follow-up to the MvLR for estimating linear association between predictors and responses. In terms of the statistical calculus, there is no difference between all variables, including both DVs and IVs in the model; therefore, it is possible to use the model to compare variables. However, this model cannot be used as a prediction model.

In addition, when the number of DVs and IVs is one, the analytical model becomes the Deming regression, which is a method widely used for the comparative evaluation of equipment in chemistry.

References

1. McArdle, B.H., Anderson, M.J.: Fitting multivariate models to community data: a comment on distance based redundancy analysis. Ecology **82**, 290–297 (2001)
2. Elo, S., Kyng, H.: The qualitative content analysis process. J. Adv. Nurs. **62**, 107–115 (2008)
3. Bartlett, M.S.: An inverse matrix adjustment arising in discriminant analysis. Ann. Math. Stat. **22**, 107–111 (1951)
4. Tatsuoka, M.M., Lohnes, P.R.: Multivariate Analysis: Techniques for Educational and Psychological Research. Macmillan Publishing Co, Inc., New York (1988)
5. Lutz, J.G., Eckert, T.L.: The relationship between canonical correlation analysis and multivariate multiple regression. Educ. Psychol. Meas. **54**, 666–675 (1994)
6. Dattalo, P.: Analysis of Multiple Dependent Variables. Oxford University Press, New York (2013)
7. Wold, H.: Path models with latent variables: the NIPALS approach. In: Quantitative Sociology, pp 307–357. Elsevier (1975)
8. Rohlf, F.J., Corti, M.: Use of two-block partial least-squares to study covariation in shape. Syst. Biol. **49**, 740–753 (2000)
9. Wold, S., Ruhe, A., Wold, H., Dunn III, W.J.: The collinearity problem in linear regression. The partial least squares (PLS) approach to generalized inverses. SIAM J. Sci. Stat. Comput. **5**, 735–743 (1984)
10. Breiman, L., Friedman, J.H.: Predicting multivariate responses in multiple linear regression. J. Roy. Stat. Soc. Ser. B (Stat. Methodol.) **59**, 3–54 (1997)
11. Yuan, M., Ekici, A., Lu, Z., Monteiro, R.: Dimension reduction and coefficient estimation in multivariate linear regression. J. Roy. Stat. Soc. Ser. B (Stat. Methodol.) **69**, 329–346 (2007)
12. Reinsel, G.C., Velu, R.P.: Multivariate linear regression. In: Multivariate Reduced-Rank Regression, pp 1–14. Springer (1998)
13. Harrell Jr., F.E.: Regression Modeling Strategies: With Applications to Linear Models, Logistic and Ordinal Regression, and Survival Analysis. Springer, Cham (2015)
14. Everitt, B.S., Dunn, G.: Applied Multivariate Data Analysis, vol. 2. Wiley Online Library, Chichester (2001)

15. Velu, R., Reinsel, G.C.: Multivariate Reduced-Rank Regression: Theory and Applications, vol. 136. Springer Science and Business Media, Heidelberg (2013)
16. Berkson, J.: Are there two regressions? J. Am. Stat. Assoc. **45**, 164–180 (1950)
17. Durbin, J.: Errors in variables. Revue de l'institut International de Statistique **22**, 23–32 (1954)
18. Zamar, R.H.: Robust estimation in the errors-in-variables model. Biometrika **76**, 149–160 (1989)
19. Maronna, R.: Principal components and orthogonal regression based on robust scales. Technometrics **47**, 264–273 (2005)
20. Golub, G.H., Reinsch, C.: Singular value decomposition and least squares solutions. Numerische Mathematik **14**, 403–420 (1970)
21. Brown, M.L.: Robust line estimation with errors in both variables. J. Am. Stat. Assoc. **77**, 71–79 (1982)
22. Golub, G.H., Van Loan, C.F.: An analysis of the total least squares problem. SIAM J. Numer. Anal. **17**, 883–893 (1980)
23. Wang, J., et al.: Package "robust" (2017)

Supervised Multiclass Classifier
for Autocoding Based on Partition Coefficient

Yukako Toko[1]([⊠]) [iD], Kazumi Wada[1] [iD], Shinya Iijima[1] [iD],
and Mika Sato-Ilic[1,2] [iD]

[1] National Statistics Center, 19-1 Wakamatsu-cho, Shinjuku-ku,
Tokyo 162-8668, Japan
ytoko@nstac.go.jp
[2] Faculty of Engineering, Information and Systems, University of Tsukuba,
Tsukuba, Ibaraki 305-8573, Japan

Abstract. The classification of objects based on classification codes is an important task for data processing in the field of official statistics. In our previous study, the supervised multiclass classifier was developed for autocoding, which has the advantages of simplicity and practical calculation time. However, the previous algorithm classified a few objects incorrectly. To address this problem, a new supervised multiclass classifier is proposed that extends the previously proposed classifier algorithm by applying the idea of partition coefficient or partition entropy. Numerical evaluation shows that the proposed algorithm has a better performance as compared to the previously proposed algorithm.

Keywords: Text classification · Coding · Supervised classification
Partition coefficient · Partition entropy

1   Introduction

This paper presents a novel multiclass classifier algorithm for autocoding based on the idea of partition coefficient or partition entropy (Bezdek 1981; Bezdek et al. 1999).

In the previous study, the supervised multiclass classifier was developed for autocoding based on a simple machine learning technique (Toko et al. 2017; Tsubaki et al. 2017; Shimono et al. 2018). Originally, the multiclass classifier algorithm was developed to improve the efficiency of a coding task that classifies objects according to their corresponding classification codes. This task, often required for survey data processing in the field of official statistics, was traditionally done manually by humans engaged in coding, with their specialized knowledge of the classification. However, with improvements in computer technology, the use of autocoding has been explored in the field of official statistics. For example, Hacking and Willenborg (2012) introduced a technique to implement coding tasks for governmental surveys in the Netherlands, and demonstrated an autocoding system. Gweon et al. (2017) presented methods for automated occupation coding based on statistical learning.

© Springer International Publishing AG, part of Springer Nature 2019
I. Czarnowski et al. (Eds.): KES-IDT 2018, SIST 97, pp. 54–64, 2019.
https://doi.org/10.1007/978-3-319-92028-3_6

In the previous study, the classifier was developed for autocoding using machine learning, which had the advantages of simplicity and practical calculation time. However, this algorithm yielded a certain volume of unmatched output. To address this issue and improve the classification accuracy, a new algorithm is proposed that extends the algorithm of the previously proposed classifier by introducing the idea of partition coefficient or partition entropy. The newly developed system takes advantage of the previously developed algorithm. After implementing the training and evaluation step of the previous algorithm, the newly proposed algorithm learns iteratively to improve the classification results by considering the partition coefficient or partition entropy for each feature. This improved the classification accuracy of the algorithm. In addition, a numerical example is illustrated to clarify the efficiency of the new algorithm.

The rest of this paper is organized as follows. The previously developed autocoding system is introduced in Sect. 2. A general explanation of the partition coefficient and partition entropy is given in Sect. 3, and a novel algorithm based on partition coefficient or partition entropy is proposed in Sect. 4. The experiments and results are described in Sect. 5. Conclusions and suggestions for future work are presented in Sect. 6.

2 Previously Developed Autocoding System

The previously developed autocoding system comprised training and classification processes. The training process involved feature extraction and creation of a feature frequency table for the extracted features along using the given classification codes. In this study, the object is considered to be a short textual description. As in the previous study (Toko et al. 2017; Shimono et al. 2018), the word-level N-gram model is employed for feature extraction. Previously, word-level N-grams ($N = 1, 2, 3, \ldots$) were taken from the word sequences of text descriptions after tokenizing each description by MeCab (Kudo et al. 2004), a dictionary-attached morphological analyser. Here, 1-grams (any word) and 2-grams (any sequence of two consecutive words) are taken as features. The system implements the following procedures in the classification process.

(Step 1) The system extracts features in the same manner as the training system.
(Step 2) Each extracted feature refers to the feature frequency table provided by the training process for retrieval of the corresponding classification codes and frequency.
(Step 3) The system determines the most promising class to assign a classification code by the following idea (Toko et al. 2017; Tsubaki et al. 2017; Shimono et al. 2018). Let multinomial classes C take values in $\{1, \ldots, K\}$, and let $\mathbf{F} = (F_1, \ldots, F_J)$ be a J-dimensional random variable whose elements take a value of 0 or 1, which respectively indicate the absence or presence of a particular feature. Then, as each feature is assumed to be conditionally independent of any other feature given C, the conditional probability of the features of $\mathbf{F}$ for a given class C can be written as

$$P\left(F_j = f_j, j = 1, \ldots, J | C = k\right) = \prod_{j=1}^{J} P\left(F_j = f_j | C = k\right)$$

$$= \prod_{j=1}^{J} p_{jk}^{f_j} \left(1 - p_{jk}\right)^{1-f_j}, \tag{1}$$

where $p_{jk} = P(F_j = 1 | C = k)$ for $k = 1, \ldots, K$.

Let n_k be the number of objects in a class k in the training dataset, and n_{jk} be the number of objects in a class k in the training dataset with $f_j = 1$. The maximum likelihood estimate of p_{jk} can be written as

$$\hat{p}_{jk} = \frac{n_{jk} + \beta}{n_k + \alpha}. \tag{2}$$

α and β are parameters that prevent $\hat{p}_{jk}$ from being equal to 0 or 1.

Under the assumption that $P(C = k) = p_k$, the posterior probability $P(C = k | F_j = f_j, j = 1, \ldots, J)$ is proportional to

$$p_k \prod_{j=1}^{J} p_{jk}^{f_j} \left(1 - p_{jk}\right)^{1-f_j}. \tag{3}$$

Under the sensitivities $\beta_{jk} = \log \frac{p_{jk}}{1-p_{jk}}$, the posterior probabilities have been described as follows (Tsubaki et al. 2017):

$$P\left(C = k | F_j = f_j, j = 1, \ldots, J\right) = \frac{exp\left(\sum_{j=1}^{J} f_j \beta_{jk}\right)}{\sum_{l=1}^{K} exp\left(\sum_{j=1}^{J} f_j \beta_{jl}\right)}, \tag{4}$$

when $p_k = a\left\{\prod_{j=1}^{J} \left(1 - p_{jk}\right)\right\}^{-1}, a \neq 0, \forall k$ are assumed. Equation (4) can be rewritten as follows:

$$P\left(C = k | F_j = f_j, j = 1, \ldots, J\right) = \frac{p_k \prod_{j=1}^{J} p_{jk}}{\sum_{l=1}^{K} \left(p_l \prod_{j=1}^{J} p_{jl}\right)}, \tag{5}$$

when $f_j = 1$. From (5), it can be seen that $P(C = k | F_j = f_j, j = 1, \ldots, J)$ when $f_j = 1$ is influenced only by n_{jk}. $\tilde{p}_{jk}$ was defined as follows (Toko et al. 2017; Shimono et al. 2018):

$$\tilde{p}_{jk} = \frac{n_{jk} + \beta}{n_j + \alpha}, n_j = \sum_{k=1}^{K} n_{jk}. \tag{6}$$

Assignment of the classification code is performed based on this $\tilde{p}_{jk}$. In the previous study, $\alpha = -0.111111$ and $\beta = -0.444444$ were set heuristically.

3 Partition Coefficient and Partition Entropy

Partition coefficient is a validity index associated with fuzzy c-means method (Bezdek 1981). The partition coefficient is defined as

$$PC(K) = \frac{1}{J} \sum_{k=1}^{K} \sum_{j=1}^{J} \mu_{kj}^2, \tag{7}$$

where μ_{kj} indicates the degree of belongingness of an object j to a cluster k under the conditions

$$\mu_{kj} \in [0, 1], \quad \sum_{k=1}^{K} \mu_{kj} = 1, \quad j = 1, \ldots, J. \tag{8}$$

From (7) and (8), $PC(K)$ satisfies $\frac{1}{K} \leq PC(K) \leq 1$. This value shows the status of clustering. Specifically, if the result of clustering shows clearer clustering in which the degree of belongingness is close to 0 or 1, then $PC(K)$ has a larger value. Similarly, if the result of clustering is uncertain, that is, almost all values of μ_{kj} are close to $1/K$, then $PC(K)$ has a smaller value.

Furthermore, the partition entropy is defined as follows:

$$PE(K) = -\frac{1}{J} \sum_{k=1}^{K} \sum_{j=1}^{J} \mu_{kj} \log_2 \mu_{kj}. \tag{9}$$

From (8) and (9), $PE(K)$ satisfies $0 \leq PE(K) \leq \log_2 K$. As this value shows the status of clustering measured based on entropy, a clearer clustering has a smaller value of $PE(K)$.

4 Proposed Algorithm Based on Partition Coefficient

A novel algorithm is proposed for an autocoding procedure based on partition coefficient. From the definition of $\tilde{p}_{jk}$ shown in (6), $\tilde{p}_{jk}$ satisfies the following conditions when $\alpha = \beta = 0$:

$$\tilde{p}_{jk} \in [0, 1], \quad \sum_{k=1}^{K} \tilde{p}_{jk} = 1, \quad j = 1, \ldots, J. \tag{10}$$

Therefore, if $\widetilde{PC}(K)$ is defined as

$$\widetilde{PC}(K) = \frac{1}{J}\sum_{k=1}^{K}\sum_{j=1}^{J}\tilde{p}_{kj}^2, \tag{11}$$

then, from (10) and (11), $\widetilde{PC}(K)$ satisfies $\frac{1}{K} \leq \widetilde{PC}(K) \leq 1$. In the same manner, if $\widetilde{PE}(K)$ is defined as

$$\widetilde{PE}(K) = -\frac{1}{J}\sum_{k=1}^{K}\sum_{j=1}^{J}\tilde{p}_{kj}\log_2\tilde{p}_{kj}, \tag{12}$$

then, from (10) and (12), $\widetilde{PE}(K)$ satisfies $0 \leq \widetilde{PE}(K) \leq \log_2 K$. If $\widetilde{PC}(K)$ and $\widetilde{PE}(K)$ for each feature j are considered, then the following PC_j and PE_j can be considered.

$$PC_j = \sum_{k=1}^{K}\tilde{p}_{jk}^2, j = 1,\ldots,J \tag{13}$$

$$PE_j = -\sum_{k=1}^{K}\left(\tilde{p}_{jk}\log_2\tilde{p}_{jk}\right), j = 1,\ldots,J \tag{14}$$

From (10), (13), and (14), PC_j and PE_j satisfy the following conditions:

$$\frac{1}{K} \leq PC_j \leq 1, \ j = 1,\ldots,J, \tag{15}$$

$$0 \leq PE_j \leq \log_2 K, \ j = 1,\ldots,J. \tag{16}$$

The value of PC_j shows the status of classification of the j-th feature over K classification codes. The clearer classification of each feature has a larger value of PC_j. The value of PE_j also shows the status of classification of the j–th feature over K classification codes. However, as it is an index of entropy, the clearer classification of each feature has a smaller value of PE_j.

The main purpose of the proposed algorithm is to address the shortcoming of the previously proposed coding system that exhibits a certain volume of unmatched output. For this purpose, the property of partition coefficient and partition entropy for each feature shown in (13), (14), (15), and (16) are utilized for extracting clearly classified features from the unmatched features.

Based on the partition coefficient and partition entropy for each feature shown in (13) and (14), the following algorithm is proposed.

(Step 1) The system assigns classification codes in the same manner as our previous system does. However, $\alpha = \beta = 0$ is set in the present study.

(Step 2)   The system creates a partition coefficient table and partition entropy table for each feature in the feature frequency table along with the most promising classification code. First, the system determines the most promising classification code for each feature in the feature frequency table based on the frequencies in the feature frequency table. Second, from (13), it calculates the partition coefficient for each feature. In the same manner, from (14), it calculates the partition entropy for each feature.

Finally, the calculated values are tabulated in a partition coefficient table or partition entropy table. However, the system does not tabulate these when $n_j = 1$ to respectively prevent PC_j from being equal to 1 and PE_j from being equal to 0 with just one trained object.

(Step 3)   The system re-assigns a classification code to each of the unmatched object based on the values of PC_j or PE_j. It gives a classification code that has a maximum value of PC_j in the partition coefficient table. Also, it gives a classification code that has a minimum value of PE_j in the partition entropy table.

(Step 4)   The threshold for PC_j is determined using the following steps. First, the system extracts unmatched objects from the re-assigned evaluated data. Then, it plots PC_j for each unmatched object. Finally, the threshold is heuristically determined. In the same manner, the threshold for PE_j is also determined.

(Step 5)   The system re-assigns a classification code to each of the unmatched objects based on the values of PC_j in the partition coefficient table. First, the system removes data that have the values of PC_j larger than the threshold determined in Step 4 from the partition coefficient table. Then, it re-assigns a classification code in the same manner as Step 3 using the remaining data in the partition coefficient table. Also, the system re-assigns a classification code based on the values of PE_j in the partition entropy table. First the system removes data that have the values of PE_j smaller than the threshold determined in Step 4 from the partition entropy table. Then, it re-assigns in the same manner as Step 3 using the remaining data in the partition entropy table.

(Step 6)   Steps 4 and 5 are iterated until the re-assigned classification accuracy is below a heuristically determined threshold. In this study, the threshold value is set as 0.1.

5   Experiment and Results

For performance evaluation, the Stack Overflow dataset is used, a publicly available short description dataset published by Xu et al. (2015) in Kaggle. The dataset contains 20,000 objects, consisting of question titles in English and 20 different classification codes (see Table 1). The following pre-processing was performed to prepare for the experiments. First, all letters in the question titles were converted to lowercase to achieve efficient feature extraction. Second, symbols that were clearly unnecessary for classification were removed from the question titles. Finally, the dataset was randomly divided into 18,000 objects for training and 2,000 objects for evaluation. This division task was repeated three times to prepare three pairs of datasets (dataset 1, 2, 3) for training and evaluation.

Table 1. Overview of the stack overflow dataset

No.	Class	Classification code	Number of objects in the dataset	No.	Class	Classification code	Number of objects in the dataset
1	wordpress	1	1,000	11	spring	11	1,000
2	oracle	2	1,000	12	hibernate	12	1,000
3	svn	3	1,000	13	scala	13	1,000
4	apache	4	1,000	14	sharepoint	14	1,000
5	excel	5	1,000	15	ajax	15	1,000
6	matlab	6	1,000	16	qt	16	1,000
7	visual-studio	7	1,000	17	drupal	17	1,000
8	cocoa	8	1,000	18	linq	18	1,000
9	osx	9	1,000	19	haskell	19	1,000
10	bash	10	1,000	20	magento	20	1,000

The proposed algorithm assigned a classification code to each object (or feature) of the evaluation dataset. Table 2 shows the classification accuracy of dataset 1 assigned based on the partition coefficient. Let N be the number of total input objects (or features), N_t be the number of input objects for the t-th evaluation, and M_t be the number of matched objects in the t-th evaluation. Then, the "Accuracy" and "Total accuracy" metrics in Table 2 are defined as follows:

$$\text{Accuracy} = \frac{M_t}{N_t}, \quad \text{Total accuracy} = \frac{\sum_{i=1}^{T} M_i}{N}.$$

After assigning classification codes using the previously developed classifier, classification codes were reassigned to the unmatched objects (or features) based on the partition coefficient. After the first evaluation based on partition coefficient (first re-assignment), we plotted values of the partition coefficient of each unmatched object (or feature) by sorting the values in descending order (see Fig. 1). From the plotted figure, a threshold γ for the partition coefficient was heuristically set for the next evaluation. In this experiment, $\gamma = 1$ was set for the second evaluation since it was considered as an inflection point. In the same manner, from Fig. 2, $\gamma = 0.5$ was set for the third evaluation.

From Table 2, it is observed that the total classification accuracy improved with each evaluation. However, the improvement rate of the accuracy gradually decreased.

Table 2. Classification accuracy assigned based on the PC (partition coefficient)

	Number of evaluated objects	Number of matched objects	Accuracy	Total accuracy
Evaluation by previous classifier	2,000	1,304	0.652	0.652
First evaluation by PC	696	458	0.658	0.881
Second evaluation by PC	238	52	0.218	0.907
Third evaluation by PC	186	22	0.118	**0.918**

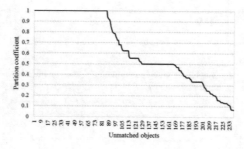

Fig. 1. The values of partition coefficient of each unmatched object (or feature) after first evaluation by partition coefficient.

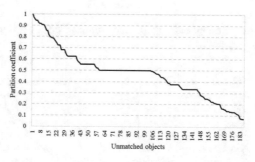

Fig. 2. The values of partition coefficient of each unmatched object (or feature) after second evaluation by partition coefficient.

The algorithm based on partition entropy was also applied to the same dataset (see Table 3). Figures 3 and 4 show the values of the partition entropy of each unmatched object (or feature) after each evaluation. These figures were plotted by sorting the values in ascending order. As in the algorithm based on the partition coefficient, each value of threshold γ was heuristically set from Figs. 3 and 4. In this experiment, $\gamma = 0$ was set for the second evaluation and $\gamma = 1$ for the third evaluation. Table 3 shows the classification accuracy assigned based on the partition entropy. The total classification accuracy improves with each evaluation as in the case of the algorithm based on the partition coefficient. Furthermore, the final performances of both proposed algorithms were virtually the same.

Table 3. Classification accuracy assigned based on the PE (partition entropy)

	Number of evaluated objects	Number of matched objects	Accuracy	Total accuracy
Evaluation by previous classifier	2,000	1,304	0.652	0.652
First evaluation by PE	696	453	0.651	0.879
Second evaluation by PE	243	54	0.222	0.906
Third evaluation by PE	189	22	0.116	**0.917**

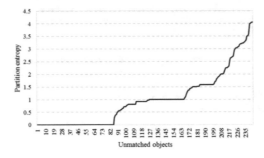

Fig. 3. The values of partition entropy of each unmatched object (or feature) after first evaluation by partition entropy.

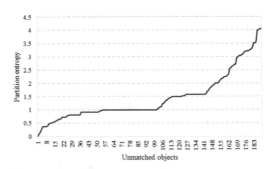

Fig. 4. The values of partition entropy of each unmatched object (or feature) after second evaluation by partition entropy.

Also the performance of the proposed algorithm was compared with that of the original classifier developed in the previous study. Table 4 lists the classification accuracy of each algorithm. For assigning classification codes using the original classifier, $\alpha = -0.111111$ and $\beta = -0.444444$ were set for calculation of $\tilde{p}_{jk}$ as in the previous study. On the other hand, for assigning classification codes with the proposed algorithm, $\alpha = \beta = 0$ was set as illustrated in Sect. 4. From the results of Table 4, it is observed that the classification accuracy of the algorithms based on the partition coefficient and partition entropy is better than that of the original classifier. In addition, it has been known that the performance of the original classifier is better than that of the random forest model and the CART model (Toko et al. 2017).

Table 4. Comparison of classification accuracy obtained from the algorithms proposed in this work and proposed previously

	Number of evaluated objects	Number of total matched objects			Total accuracy		
		Original classifier	Partition coefficient	Partition entropy	Original classifier	Partition coefficient	Partition entropy
Dataset 1	2,000	1,795	1,836	1,833	0.898	**0.918**	0.917
Dataset 2	2,000	1,799	1,833	1,839	0.900	0.917	**0.920**
Dataset 3	2,000	1,773	1,841	1,837	0.887	**0.921**	0.919

6 Conclusions

This paper presented a novel multiclass classification algorithm based on the partition coefficient or partition entropy. To address a flaw in our previously proposed multiclass classifier, which allowed a certain volume of unmatched objects (or features) to remain, the idea of partition coefficient or partition entropy was applied. The objective was to use the extra information on the classification status of each object (or feature) to extract clearly classified unmatched objects (or features). The remaining objects (or features) were applied to the classifier iteratively. Numerical evaluation showed that the proposed algorithm improved the performance of the multiclass classifier from our previous study. In future research, methods that determine thresholds for the values of partition coefficient and partition entropy must be considered for selecting clearly classified objects (or features). In addition, numerical investigation based on comparison with other machine learning methods is also required.

Acknowledgements. We would like to thank Kaggle for making the Stack Overflow dataset available.

References

Bezdek, J.C.: Pattern Recognition with Fuzzy Objective Function Algorithms. Plenum Press, New York (1981)

Bezdek, J.C., Keller J., Krisnapuram, R., Pal, N.R.: Fuzzy Models and Algorithms for Pattern Recognition and Image Processing. Kluwer Academic Publishers, Dordrecht (1999)

Gweon, H., Schonlau, M., Kaczmirek, L., Blohm, M., Steiner, S.: Three methods for occupation coding based on statistical learning. J. Off. Stat. **33**(1), 101–122 (2017)

Hacking, W., Willenborg, L.: Method series theme: coding; interpreting short descriptions using a classification. In: Statistics Methods. Statistics Netherlands (2012). https://www.cbs.nl/en-gb/our-services/methods/statistical-methods/throughput/throughput/coding. Accessed 16 Jan 2018

Kudo, T., Yamamoto, K., Matsumoto, Y.: Applying conditional random fields to Japanese morphological analysis. In: The 2004 Conference on Empirical Methods in Natural Language Processing on Proceedings, Barcelona, Spain, pp. 230–237 (2004)

Shimono, T., Wada, K., Toko, Y.: A supervised multiclass classifier using machine learning algorithm for autocoding. Res. Mem. Off. Stat. **75**, 41–60 (2018). (in Japanese)

Toko, Y., Wada, K., Kawano, M.: A supervised multiclass classifier for an autocoding system. J. Rom. Stat. Rev. **4**, 29–39 (2017)

Tsubaki, H., Wada, K., Toko, T.: An extension of Taguchi's T method and standardized misclassification rate for supervised classification with only binary inputs. In: Proceedings of the ANQ Congress, Kathmandu, Nepal (2017)

Xu, J., Wang, P., Tian, G., Xu, B., Zhao, J., Wang, F., Hao, H.: Short text clustering via convolutional neural networks. In: NAACL-HLT on Proceedings, Denver, Colorado, USA, pp. 62–69 (2015)

An Empirical Analysis of Volatility by the SIML Estimation with High-Frequency Trades and Quotes

Hiroumi Misaki[✉] [iD]

University of Tsukuba, Tennodai 1-1 -1, Tsukuba, Ibaraki 305-8577, Japan
hmisaki@risk.tsukuba.ac.jp

Abstract. Estimating the volatility of financial asset prices is an issue of considerable importance in financial econometrics research. For estimating the integrated volatility by using high frequency data, Kunitomo and Sato [6] proposed the separating information maximum likelihood (SIML) method for estimating volatility from data contaminated by market microstructure noise. Subsequently, Misaki and Kunitomo [10] investigated the method when used with randomly sampled data. The method has not been tested for estimating the volatility of individual stocks with high-frequency data. This article analyzes tick-by-tick prices data to compare daily volatility estimates calculated with SIML estimation with estimates delivered by conventional methods. We first test the method with transaction prices from several firms and then with quote and transaction prices to investigate its robustness against noise. Our findings suggest that SIML estimation is useful for analyzing actual markets with high-frequency data.

Keywords: Financial econometrics · Integrated volatility
Market microstructure noise · High-frequency data · SIML
Random sampling

1 Introduction

Estimating the volatility of financial asset prices is a significant research goal in finance, since volatility has important effects on option pricing, asset allocation, risk management, and other practical concerns. Over the last decade, high-frequency financial data, or *tick-by-tick* data, has become available. High-frequency asset-price data has made estimations of daily volatility possible, and researchers continue to investigate the estimation problem.

Integrated volatility is a natural measure of volatility for this purpose. *Realized volatility* is conventionally used to estimate integrated volatility [2]. However, realized volatility is known to be sensitive to the microstructure noise that contaminates actual market data. Several alternative estimation methods have been developed to address this problem (e.g. [1,3,9,11]).

In this respect Kunitomo and Sato [6] proposed a statistical method called the separating information maximum likelihood (SIML) method for estimating

© Springer International Publishing AG, part of Springer Nature 2019
I. Czarnowski et al. (Eds.): KES-IDT 2018, SIST 97, pp. 65–75, 2019.
https://doi.org/10.1007/978-3-319-92028-3_7

integrated volatility and integrated covariance from high-frequency data with market microstructure noise. The SIML estimator has been shown to have reasonable asymptotic properties and finite sample properties [7,8].

The SIML method was originally defined in terms of equidistant observations, but transactions occur at random intervals in actual markets. Misaki and Kunitomo [10] investigated the properties of the SIML estimation with data that is randomly sampled and includes market microstructure noise. Then Kunitomo et al. [5] studied a multivariate case and reported reasonable results. Both studies are based on asymptotic analysis and massive Monte Carlo simulations.

In this paper, we analyze high-frequency financial data from the Japanese stock market. The analysis uses tick-by-tick transaction prices from the Osaka Securities Exchange (OSE) and transaction and quote prices from the Tokyo Stock Exchange (TSE)[1]. The SIML estimator is applied to this data to estimate daily volatility, and we compare this estimation with several alternative estimators.

This study has both econometric and practical significance since the SIML estimation has not ever been applied to individual stock data. Thus, this is the first empirical study investigating the effectiveness of the SIML estimator when applied to individual stock data.

Evaluating the accuracy of high-frequency volatility estimators is complicated because integrated volatility is unobservable. In the latter half of our empirical analysis, we shall assess the accuracy of the SIML estimator using estimations from quote prices along with transaction prices. This comparison demonstrates how robust the SIML method is against market microstructure noise, as quote and transaction price data are contaminated with distinct types of noise.

The structure of the paper is as follows. Section 2 summarizes the general formulation under which Misaki and Kunitomo [10] investigated the properties of the SIML estimator. Section 3 investigates the SIML estimator of integrated volatility with high-frequency transaction data from 10 firms. This section compares the SIML with alternative estimators, and is followed by a more detailed analysis with trade and quote data in Sect. 4. Conclusions are given and discussed in Sect. 5.

2   Estimating Volatility with High-Frequency Data

In this section, we briefly introduce the SIML estimator as applied to randomly sampled univariate data with market microstructure noise, based on Misaki and Kunitomo [10]. Several assumptions were omitted to save space. For a more precise description that considers asymptotic properties and simulation results including multivariate case, see [5–8, 10].

---

[1] Now both markets are integrated under the Japan Exchange Group, Inc.

2.1   General Formulation

Let $y(t_i^n)$ be the i−th observation of the logarithmic (log) price at t_i^n for $0 = t_0^n < t_1^n < \cdots < t_{n*}^n \le t_n^n = 1$ and $t_{n*}^n = \max_{t_i^n \le 1} t_i^n$. n and n^* are indices of the sample size; n^* is the stochastic index while n is the constant index.

We assume that the underlying continuous process $X(t)$ $(0 \le t \le 1)$ is not necessarily the same as the observed log price at $t_i^n (i = 1, \cdots, n^*)$ and that

$$X(t) = X(0) + \int_0^t \sigma_x(s)dB(s) \ (0 \le t \le 1), \tag{1}$$

where $B(s)$ represents standard Brownian motion and $\sigma_x(s)$ is the instantaneous volatility function. The main statistical goal is to estimate the integrated volatility

$$\sigma_x^2 = \int_0^1 \sigma_x^2(s)ds \tag{2}$$

of the underlying continuous process $X(t)$ $(0 \le t \le 1)$ from the set of discretely observed prices $y(t_i^n)$ that are generated by $y(t_i^n) = h\left(X(t_i^n), y(t_{i-1}^n), u(t_i^n)\right)$, where $u(t_i^n)$ is market microstructure noise. The time interval $[0, 1]$ typically corresponds to one day, in practice.

2.2   The SIML Method

First we consider equidistant observation intervals $h_n = y_{t_i^n} - y_{t_{i-1}^n} = 1/n$ and noise with a simple structure $y(t_i^n) = X(t_i^n) + u(t_i^n)$. Assuming that $u(t_i^n)$ are independently, identically and normally distributed as $N(0, \sigma_u^2)$ given the initial condition y_0, we have

$$\mathbf{y}_n \sim N_n\left(y(0)\mathbf{1}_n, \sigma_u^2\mathbf{I}_n + h_n\sigma_x^2\mathbf{C}_n\mathbf{C}_n'\right), \tag{3}$$

where the C_n is an $n \times n$ matrix

$$\mathbf{C}_n = \begin{pmatrix} 1 & 0 & \cdots & 0 & 0 \\ 1 & 1 & 0 & \cdots & 0 \\ 1 & 1 & 1 & \cdots & 0 \\ 1 & \cdots & 1 & 1 & 0 \\ 1 & \cdots & 1 & 1 & 1 \end{pmatrix} = \begin{pmatrix} 1 & 0 & \cdots & 0 & 0 \\ -1 & 1 & 0 & \cdots & 0 \\ 0 & \ddots & \ddots & \ddots & 0 \\ 0 & \cdots & -1 & 1 & 0 \\ 0 & \cdots & 0 & -1 & 1 \end{pmatrix}^{-1} \tag{4}$$

and the vector $\mathbf{1}_n' = (1, \cdots, 1)$. When the observations are randomly sampled, we consider n^* instead of n. Then $\mathbf{y}_{n*}$ is transformed into $\mathbf{z}_{n*}(= (z_1, \ldots, z_{n*})')$ as follows:

$$\mathbf{z}_{n*} = \sqrt{n^*} \ \mathbf{P}_{n*} \ \mathbf{C}_{n*}^{-1} (\mathbf{y}_{n*} - \bar{\mathbf{y}}_0) \tag{5}$$

where $\bar{\mathbf{y}}_0 = y(0)\mathbf{1}_{n*}$, $\mathbf{P}_{n*} = (p_{jk})$ and for $j, k =, \ldots, n^*$,

$$p_{jk} = \sqrt{\frac{2}{n^* + \frac{1}{2}}} \cos\left[\frac{2\pi}{2n^* + 1}\left(j - \frac{1}{2}\right)\left(k - \frac{1}{2}\right)\right]. \tag{6}$$

Then the transformed variables z_k $(k = 1, \cdots, n^*)$ given n^* are mutually independent and $z_k \sim N(0, \sigma_x^2 + a_{k,n^*}\sigma_u^2)$, where

$$a_{k,n^*} = 4n^* \sin^2\left[\frac{\pi}{2}\left(\frac{2k-1}{2n^*+1}\right)\right] \ (k = 1, \cdots, n^*) . \tag{7}$$

Let m and l be positive integers dependent on n^*, hence we write m_{n^*} and l_{n^*}. Then we define the SIML estimators of $\hat{\sigma}_x^2$ and $\hat{\sigma}_u^2$ as

$$\hat{\sigma}_x^2 = \frac{1}{m_{n^*}}\sum_{k=1}^{m_{n^*}} z_k^2 \quad \text{and} \quad \hat{\sigma}_u^2 = \frac{1}{l_n}\sum_{k=n^*+1-l_{n^*}}^{n^*} a_{k,n^*}^{-1} z_k^2 , \tag{8}$$

respectively. The numbers of terms m_{n^*} and l_{n^*} are dependent on n^* so that $m_{n^*}, l_{n^*} \to \infty$ as $n \to \infty$. We impose order requirements such that $m_{n^*} = O_p(n^\alpha)$ $(0 < \alpha < \frac{1}{2})$ and $l_{n^*} = O_p(n^\beta)$ $(0 < \beta < 1)$ for σ_x^2 and σ_u^2, respectively. In the following empirical analysis we set $\alpha = 0.4$ and $\beta = 0.8$.

3   Empirical Analysis

3.1   High-Frequency Data from the OSE

We focus on the following 10 assets, which are the 10 most-frequently traded on the OSE: Nintendo Co., Ltd. (7974), Murata Manufacturing Co., Ltd. (6981), NIDEC Corp. (6594), ROHM Co., Ltd. (6963), Omron Corp. (6645), Funai Electric Co., Ltd. (6839), Mori Seiki Co., Ltd. (6141), Ono Pharmaceutical Co., Ltd. (4528), Benesse Holdings, Inc. (9783) and Sumitomo Forestry Co., Ltd. (1911). The numbers in parentheses denote the security code.

The sample period ranges from March 2nd, 2006 to May 27th, 2008, containing 546 business days. Throughout this period, the morning session of the OSE opened at 9:00 and closed at 11:00, and the afternoon session ran from 12:30 to 15:10.

We omit from the analysis days on which only the one of the sessions opens, to keep the time scale even. If multiple transactions have the same time stamp, we select the first observation and omit the others. This data-cleaning process might seem irrational, but Barndorff-Nielsen et al. [4] studied sensitivity to data-cleaning methods and found that the estimates are not sensitive to treatments of the data recorded at the same time. So we employ the simple method for the ease of computations in the analysis discussed below.

3.2   Estimation of Integrated Volatility

The main purpose is to estimate integrated volatility by using the tick-by-tick data of each asset. We investigate the properties of the SIML estimator and several alternative estimators for integrated volatility.

The SIML estimate for the i-th asset on day s is denoted by $V_{s,i}^{\text{SI}}$. It is computed based on all the transaction prices on day s, except for observations with the same time stamp.

We also compute the following estimators for the sake of comparison. (i) Realized volatility $V_{s,i}^{RV}$ as calculated from all data. (ii) Realized volatility $V_{s,i}^{\Delta}$ as calculated from evenly spaced data, where Δ represents the sampling interval, using the previous-tick method. Specifically, we set $\Delta = 1/6, 1, 5$ and 20 min. (iii) The open-to-close estimator $V_{s,i}^{OC}$, which is the square of the open-to-close return. The mean of this value over the sample period provides a rough estimate of an asset's average daily volatility.

The OSE takes a mid-day recess between the morning and afternoon sessions. In this study, we compute estimates for the morning session and the afternoon session separately, then simply add them to obtain the daily estimates, excepting the open-to-close estimator.

We summarize the results of our volatility estimates in Tables 1 and 2. In the whole of Table 1 and the upper rows in Table 2, the multipliers "E-04" are omitted to simplify the presentation. The upper row for each asset in Table 1 shows the means of the estimates over the sample days, while the lower row shows their standard deviations. The realized volatility estimates grow larger as the sampling intervals shorten, and $V_{s,i}^{RV}$ are the largest of the estimates we compared. This result is expected when market microstructure noise is present, as the literature indicates that realized volatility tends to be overestimated in this case. On the other hand, $V_{s,i}^{SI}$ gives a relatively reasonable value, when compared with $V_{s,i}^{OC}$. We also find that $V_{s,i}^{SI}$ tends to be similar to $V_{s,i}^{5}$ and $V_{s,i}^{20}$ in terms of both mean and standard deviation.

This trend is made clearer by the summary in Table 2. The upper row for each firm in Table 2 shows the biases of respective estimators, defined as the difference between the mean of the estimates and $V_{s,i}^{OC}$. The lower row for each firm lists the ratio of these biases to $V_{s,i}^{OC}$. The means of the biases and the means of the ratios for all 10 firms are listed at the bottom of the table. We find that $V_{s,i}^{SI}$ is relatively accurate along with $V_{s,i}^{20}$, while $V_{s,i}^{5}$ tends to overestimate volatility by 20% on average. Therefore, we conclude that the SIML method returns reasonable estimates when compared to the empirical realized volatilities of individual stocks.

4 Further Analysis with Trades and Quotes

4.1 Using Quote Data for Volatility Estimation

In the previous section, we argued that the SIML estimator returns values close to those returned by the open-to-close estimator. The open-to-close estimate, however, is used as a criterion for an accurate estimate assuming that the volatilities are constant throughout the observation period. On the other hand, we seek to estimate daily volatility with high-frequency data precisely because the volatility of asset prices varies day-to-day. However, the accuracy of integrated volatility estimates cannot be assessed in a straightforward manner, since true volatility cannot be measured with available data.

Data about quote prices can be used along with transaction prices to address this problem. As Barndorff-Nielsen et al. [4] note, transaction and quote prices

Table 1. Estimation of volatility of 10 assets.

Firm	SIML	RV	RV$^{1/6}$	RV^1	RV^5	RV^{20}	OC
7974	3.58	12.75	7.78	4.61	3.89	3.45	4.11
	4.34	11.63	5.95	4.39	4.62	4.52	
6981	2.42	8.29	6.39	3.79	2.73	2.44	2.44
	2.23	5.40	3.67	2.25	2.25	2.71	
6594	2.92	6.49	5.56	4.12	3.17	2.83	3.21
	2.50	3.55	2.92	2.45	2.53	2.75	
6963	2.50	4.28	3.97	3.32	2.67	2.47	2.09
	2.76	3.06	3.26	3.31	3.19	3.73	
6645	3.03	15.56	12.44	7.56	4.04	3.11	2.91
	3.57	8.96	6.53	3.52	2.92	3.19	
6839	6.30	7.70	7.52	6.94	6.09	5.45	5.84
	8.71	7.67	7.57	7.65	6.89	6.86	
6141	4.11	14.25	11.29	6.87	4.65	4.23	5.04
	4.48	11.46	6.95	3.97	4.19	5.90	
4528	2.25	4.64	4.22	3.36	2.46	2.10	1.61
	2.43	2.58	2.27	1.83	1.87	1.88	
9783	2.60	9.98	8.52	5.82	3.45	2.73	2.39
	2.85	5.92	4.43	2.79	2.47	2.76	
1911	4.86	5.23	5.26	5.06	4.73	4.25	3.59
	6.76	4.50	4.79	4.92	6.01	5.74	

are both considered noisy indicators of underlying efficient market prices. Quote and transaction prices are affected by distinct forms of market microstructure noise, so an estimator that is unaffected by both forms of noise should produce almost the same estimates from transaction and quote data. Thus, in this section, we estimate volatility from data about both transaction and quote prices, using the SIML and the realized volatility estimator. We then compare the four sets of estimates to assess the degree to which the SIML estimator is robust against market microstructure noise.

4.2   Data Description

For trade and quote analysis, we use data from Softbank Group Corp. (9984), listed on the TSE. The Softbank Group stock is one of the most traded individual stocks in Japanese financial markets in terms of both contracts and volume.

We analyzed all business days of 2016. Unlike the OSE discussed in the previous section, the TSE stock market opens from 9:00 to 11:30 and 12:30 to 15:00 on every business day. We omit one day of data because both trades and quotes were very few due to a market restriction. We then utilized 243 sample

Table 2. Biases and their ratios of volatility estimates.

Firm	SIML	RV	$RV^{1/6}$	RV^1	RV^5	RV^{20}	OC
7974	−0.54	8.64	3.67	0.49	−0.23	−0.66	0.00
	−0.13	2.10	0.89	0.12	−0.05	−0.16	0.00
6981	−0.02	5.85	3.95	1.35	0.29	0.00	0.00
	−0.01	2.40	1.62	0.55	0.12	0.00	0.00
6594	−0.29	3.28	2.35	0.91	−0.04	−0.38	0.00
	−0.09	1.02	0.73	0.28	−0.01	−0.12	0.00
6963	0.41	2.19	1.87	1.23	0.58	0.38	0.00
	0.19	1.05	0.90	0.59	0.28	0.18	0.00
6645	0.12	12.64	9.52	4.65	1.12	0.20	0.00
	0.04	4.34	3.27	1.59	0.39	0.07	0.00
6839	0.46	1.86	1.69	1.10	0.25	−0.38	0.00
	0.08	0.32	0.29	0.19	0.04	−0.07	0.00
6141	−0.93	9.21	6.25	1.83	−0.38	−0.81	0.00
	−0.18	1.83	1.24	0.36	−0.08	−0.16	0.00
4528	0.64	3.02	2.60	1.75	0.85	0.48	0.00
	0.39	1.87	1.61	1.08	0.53	0.30	0.00
9783	0.21	7.59	6.13	3.44	1.06	0.34	0.00
	0.09	3.18	2.57	1.44	0.45	0.14	0.00
1911	1.27	1.64	1.67	1.47	1.14	0.66	0.00
	0.35	0.46	0.46	0.41	0.32	0.18	0.00
Mean	0.13	5.59	3.97	1.82	0.47	−0.02	0.00
	0.07	1.86	1.36	0.66	0.20	0.04	0.00

days in total. We again use the first record if we have the same time stamp for multiple transactions or quotes, though this situation seldom occurs because this data set is reported in units of $100\,\mu s$ ($10^{-4}\,s$).

4.3 Data Analysis

We next estimate the integrated volatility for each day with the SIML and the realized volatility estimators. For the quote data, we use mid-quote prices. The set of intervals of the realized volatility and the relevant notation is almost the same as that in the previous section, except that we employ 15-min intervals instead of 20-min because we find that both 5-min and 20-min intervals produces similar statistics.

Table 3 summarizes the analysis in the same manner as Table 1. The estimate of realized volatility with tick-by-tick transaction data is larger than the other estimates, as expected. For the quote data, the overestimation is relatively moderate, but still indicates the effect of market microstructure noise. The means of SIML estimates from trade and quote prices are quite close to each other. Considering the mid-day recess, it is natural that these estimates are somewhat smaller than $V_{s,i}^{OC}$.

Table 3. Estimation of volatility: trades and quotes

Firm	SIML	RV	$RV^{1/6}$	RV^1	RV^5	RV^{15}	OC
9984	3.23	7.36	3.60	3.35	3.34	3.45	3.88
(trade)	4.15	9.91	3.76	4.28	5.61	6.57	
9984	3.30	4.63	4.85	4.99	4.18	3.38	4.01
(quote)	4.13	10.7	10.6	10.8	9.07	6.58	

We now turn to a detailed analysis of the estimates based on trade and quote data. The scatterplots in Fig. 1 plot the estimates from trade prices on the horizontal axis and the estimates from mid-quote prices on the vertical axis. Regression lines, coefficients and R-squared statistics are indicated in the plots.

The upper-left panel in Fig. 1 shows strong agreement between SIML estimates from trade and quote prices aside from a few exceptions. This strongly suggested that the SIML estimator is robust against noise. On the other hand, some values of $V_{s,i}^{RV}$ and $V_{s,i}^1$ are far above the 45° line. The fitted line of $V_{s,i}^{RV}$ shifted to the right, which shows that market microstructure noise affects transaction data more severely than it affects quote data. The plot of $V_{s,i}^{15}$ shows it produces almost the same estimates from both trades and quotes as well as $V_{s,i}^{SI}$. This is not surprising because at this time interval the trade and mid-quote prices seem to be quite close and the realized volatility, sampled at such a low frequency, is hardly affected by noise.

Throughout the research, the low-frequency realized volatilities seem to produce similar results to the SIML estimator on average. But $V_{s,i}^{15}$, for example, uses only 22 observations for each day while the SIML considers all ticks. The comprehensiveness of the SIML estimator recommends its use.

Note that we have not paid careful attention to possible outliers in the data cleaning process; we simply eliminated entries with time stamps outside the operating time and ones containing no value for prices we would use. The outlier in the plot of SIML estimates in Fig. 1 is due to a day on which a significant jump was observed in quote prices and no transactions occurred due to a market restriction. This "jump" is recorded only in quote database, so that the estimates from the transaction and quote data should differ significantly. However, except for this extreme case, the SIML estimation seems robust to the data cleaning procedure. This robustness is desirable for practical implementation because the identification of outliers is often arbitrary, and because some financial datasets do not distinguish between deals made under exceptional circumstances and those that are truly outliers.

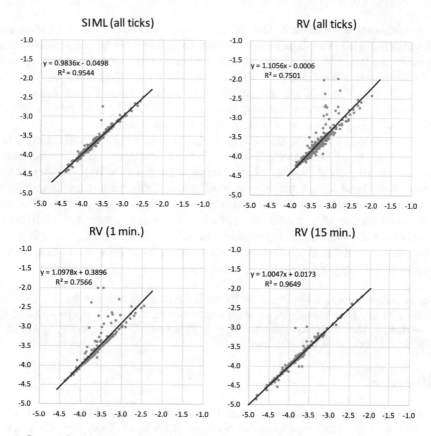

Fig. 1. Logarithmic scatterplots of the volatility estimates from transaction prices (horizontal axis) and quote prices (vertical axis). Regression lines and their coefficients are included.

5   Concluding Remarks

We tested the accuracy of the SIML method for estimating daily integrated volatility from high-frequency transaction and quote price data about individual stocks listed on the OSE and the TSE. For the sake of comparison, we also computed the conventional realized volatility estimates from the same data. These data likely include market microstructure noise.

We find that the SIML estimator provides reasonable results while most of alternatives we examined have severe biases. Since SIML estimates based on transaction prices and quote prices agree quite well, we conclude that the SIML estimator is robust against market microstructure noise. Moreover, this finding suggests that the SIML estimator is not sensitive to the data cleaning strategy employed.

In addition, we find that the SIML estimates tend to be similar to realized volatility estimates based on relatively long intervals in comparison to the summary statistics of the whole sampling period, but the estimates may differ significantly when the analysis is more detailed. In case that the estimators differ, we consider the SIML estimator to be preferable since it uses hundreds or thousands of observations, whereas the realized volatility calculated from, for instance, 15-min returns can use only 22 samples per day.

In conclusion, our analysis of high-frequency data about individual stocks suggests that the SIML method is useful for estimations of daily integrated volatility in an actual market. This finding may provide important practical implications about option pricing, asset allocation, risk management and so on.

Although we have focused on a univariate case in this paper, the SIML estimator is so simple that it can be applied to both univariate and multivariate time series with market microstructure noise. In future work, we will report on the SIML estimation of integrated covariance, correlation and other related quantities when applied to actual market data.

Acknowledgements. I would like to thank the editors and chairs for the invitation to KES-IDT-18. I also thank two anonymous reviewers for useful comments and recommendations that improved this manuscript. This research is supported by Grant for Social Science from Nomura Foundation.

References

1. Ait-Sahalia, Y., Mykland, P., Zhang, L.: How often to sample a continuous-time process in the presence of market microstructure noise. Rev. Fin. Stud. **18**(2), 351–416 (2005)
2. Andersen, T.G., Bollerslev, T., Diebold, F.X., Labys, P.: The distribution of exchange rate volatility. J. Am. Stat. Assoc. **96**, 42–55 (2001)
3. Barndorff-Nielsen, O.E., Hansen, P.R., Lunde, A., Shephard, N.: Designing realized kernels to measure the ex-post variation of equity prices in the presence of noise. Econometrica **76**(6), 1481–1536 (2008)
4. Barndorff-Nielsen, O.E., Hansen, P.R., Lunde, A., Shephard, N.: Realized kernels in practice: trades and quotes. Econ. J. **12**(3), C1–C32 (2009)
5. Kunitomo, N., Misaki, H., Sato, S.: The SIML estimation of integrated covariance and hedging coefficients with micro-market noises and random sampling. Asia Pac. Fin. Mark. **22**(3), 333–368 (2015)
6. Kunitomo, N., Sato S.: Separating information maximum likelihood estimation of realized volatility and covariance with micro-market noise. Discussion paper CIRJE-F-581, Graduate School of Economics, University of Tokyo (2008)
7. Kunitomo, N., Sato, S.: The SIML estimation of the integrated volatility of Nikkei-225 futures and hedging coefficients with micro-market noise. Math. Comput. Simul. **8**, 1272–1289 (2011)
8. Kunitomo, N., Sato, S.: Separating information maximum likelihood estimation of the integrated volatility and covariance with micro-market noise. N. Am. J. Econ. Fin. **26**, 282–309 (2013)
9. Malliavin, P., Mancino, M.: A Fourier transform method for nonparametric estimation of multivariate volatility. Ann. Stat. **37**(4), 1983–2010 (2009)

10. Misaki, H., Kunitomo, N.: On robust properties of the SIML estimation of volatility under micro-market noise and random sampling. Int. Rev. Econ. Fin. **40**, 265–281 (2015)
11. Zhang, L., Mykland, P., Ait-Sahalia, Y.: A tale of two time scales: determining integrated volatility with noisy high-frequency data. J. Am. Stat. Assoc. **100**(472), 1394–1411 (2005)

AFRYCA 3.0: An Improved Framework for Consensus Analysis in Group Decision Making

Álvaro Labella(✉) and Luis Martínez

Department of Computer Science, University of Jaén, 23071 Jaén, Spain
{alabella,martin}@ujaen.es

Abstract. Consensus reaching processes are increasingly important in group decision making problems. There are many proposals with distinct features, being complex to make comparisons and analysis between them. For this reason, AFRYCA was proposed as an analytic framework able to carry out analyses and studies in decision making problems resolution. This contribution provides new characteristics with increasing functionality for AFRYCA.

Keywords: AFRYCA · Group decision making
Consensus reaching process · Consensus model

1 Introduction

Decision Making (DM) is a common process, in which a problem with several options or alternatives can be solved by selecting the best one/s. In Group Decision Making (GDM), a set of individuals or experts with distinct points of view, are asked to find a solution to the same DM problem [4,7]. Traditionally, DM problems are solved by applying progressively alternatives solutions [1], chosen by experts [13]. Nevertheless, this process does not take into account the agreement between experts in the selection process of the solution. Because of this, Consensus Reaching Processes (CRP) emerged as an additional phase in the GDM problem resolution [14]. In a CRP, experts revise, discuss and change their preferences in order to reach a consensus and find a solution in which all of them agreed.

CRPs have become a mayor research topic within the field of GDM and a large number of proposals and approaches have been proposed for this kind of processes [2,9]. For this reason, it makes difficult to identify when a CRP is better than other in a specific GDM problem or even to set the correct configuration of the parameters of a CRP. In order to overcome this problem, *A FRamework for the analYsis of Consensus Approaches* (AFRYCA) 1.0 was proposed [9]. AFRYCA 1.0 is a *Eclipse Rich Client Platform*[1] (RCP) software tool

---

[1] http://www.eclipse.org/home/categories/rcp.php.

© Springer International Publishing AG, part of Springer Nature 2019
I. Czarnowski et al. (Eds.): KES-IDT 2018, SIST 97, pp. 76–86, 2019.
https://doi.org/10.1007/978-3-319-92028-3_8

whose functionality is provided by means of developing software components, so-called *plug-ins*, and it is used to: analyse advantages and disadvantages of consensus models, simulate CRPs for a specific GDM problem and facilitate comparisons between consensus models. AFRYCA 1.0 carries out the analysis of CRPs by simulating experts' behaviours, while imitating a real-world CRP. When AFRYCA 1.0 was used with different consensus models targeted to different contexts, some limitations were detected. Such limitations were fixed in AFRYCA 2.0 [6]. Furthermore, AFRYCA 2.0 presented multiple improvements such as new technologies, consensus models and behaviours configuration or experts' preferences visualization by means of multi-dimensional scaling. However, CRPs and consensus models are continually evolving and improving. Thus, AFRYCA 2.0 should evolve and improve at the same time. Keeping in mind this, AFRYCA 2.0 presents drawbacks such as a limited user interface, impossibility to compare several simulations each other because of the absence of the results storage, incapability to process GDM problems with different types of preference relations and multiple criteria or the necessity to know the technology in which the framework has been developed to extend or modify its functionality.

Focusing on these limitations, AFRYCA 2.0 has evolved to the version 3.0 by developing new plug-ins. The new functionalities presented in this contribution are: new user interface, storage of the results obtained in different simulations to make comparisons, capacity to deal with several types of information, new consensus models, dealing with multi-criteria GDM problems and new scripting environment integrates in AFRYCA to develop and execute scripts in distinct programming languages.

This contribution is organized as follows: Sect. 2 overviews concepts about GDM and CRP, Sect. 3 introduces the new features of AFRYCA 3.0. While, Sect. 4, provides conclusions and introduces future work.

2   Background

This section provides a review brief of the basic concepts associated with GDM and CRP.

2.1   Group Decision Making

A GDM problem is characterized by a set of experts $E = \{e_1, e_2, \ldots, e_m\}$ that provide preferences for a sequence of alternatives or options, $X = \{x_1, x_2, \ldots, x_n\}$, defined by a finite set of criteria $C = \{c_1, c_2, \ldots, c_k\}$, with the aim to find a common solution [7]. Each expert $e_i \in E$ express his/her opinion over distinct alternatives using a preference relation. A preference relation associated with the expert e_i is noted as $P_i = (p_i^{lk})_{n \times m}$ and is represented in Eq. 1 as follows:

$$P_i = \begin{pmatrix} - & \cdots & p_i^{ln} \\ \vdots & \ddots & \vdots \\ p_i^{n1} & \cdots & - \end{pmatrix} \tag{1}$$

There are several preference relations, the most common are explained below:

- *Fuzzy preference relation* [8]: In a Fuzzy Preference Relation (FPR), each valuation $p_i^{lj} = \mu_{P_i}(x_l, x_j) \in [0,1]$ represents the membership degree of e_i over x_l with respect to x_j, $l, k \in \{1, \dots, n\}, l \neq j$.
- *Linguistic preference relation* [10]: In a Linguistic Preference Relation (LPR), each valuation $p_i^{lj} \subseteq X \times X$, with membership function $\mu_p : X \times X \to S$, where S is a preestablish label set and $\mu_p(x_l, x_j) = p_{lj}$ denotes the linguistic preference degree of the alternative x_l over x_j.
- *Hesitant preference relation* [16]: In a Hesitant Preference Relation (HPR), each evaluation $p_i^{lj} \subseteq X \times X$, where $p_{lj} = \{p_{lj}^{\beta}, \beta = 1, 2, \dots, \#p_{lj}\}$ ($\#p_{lj}$ is the number of values in p_{lj}), is a hesitant fuzzy element indicating all the possible preference degree(s) of the alternative x_l over x_j.
- *Hesitant linguistic preference relation* [11,12]: In a Hesitant Linguistic Preference Relation (HLPR), each valuation $p_i^{lj} \subseteq X \times X \to S$, where S is a preestablish label set and $p_{lj} = \{p_{lj}^{\beta} \mid \beta = 1, 2, \dots, \#p_{lj}\}$ ($\#p_{lj}$ is the number of linguistic terms in p_{lj}), is a hesitant fuzzy linguistic term set indicating all the possible preference degree(s) of the alternative x_l over x_j.

Apart from these preference relations, there is another preference structure usually used in GDM, called *Decision Matrix* (DM), which is represented in Table 1. In a DM, each valuation $p_i^{lu} \subseteq X \times X \to D$, where D is any of the domains defined in the previous preference relations, shows the expert e_i opinion over the alternative x_l based on certain decision criterion c_u, unlike preference relations, that establish pairwise comparisons between alternatives.

Table 1. Decision matrix

	c_1	c_2	$\cdots$	c_k
x_1	p^{11}	p^{12}	$\cdots$	p^{1k}
x_2	p^{21}	p^{22}	$\cdots$	p^{2k}
$\vdots$	$\vdots$	$\cdots$	$\cdots$	$\vdots$
x_n	p^{n1}	p^{n2}	$\cdots$	p^{nk}

Once the experts' preferences are gathered, the classical approach to solve a GDM problem is supported by *aggregation* and *exploitation* phases [13] where:

– *Aggregation*: Experts' preferences are aggregated using an aggregation operator.
– *Exploitation*: Selecting a set of criteria, an alternative or a set of them are obtained, being such alternative/s the solution/s of the problem.

2.2 Consensus Reaching Processes

As it was aforementioned, CRP were introduced as an additional phase in GDM problem resolution, with the aim of reaching an agreement in the selection of the alternatives/s as solution/s of the problem. The consensus concept has been dealt from different points of view. Some researchers define consensus as the unanimity (almost impossible to reach in real GDM problems), others provide a more flexible definition. *Soft consensus* is a flexible consensus concept, based on the fuzzy majority, introduced by Kacprzyk [4]. Such consensus is reached when *most of the important individuals agree as to (their testimonies concerning) almost all of relevant opinions.*

CRPs are iterative and dynamic processes, in which the experts modify their initial preferences, to make their opinions closer and reach a high level of agreemet after several discussion rounds [14]. The whole process is usually guided by a *moderator* who supervises and makes suggestions to the experts involved in the process. Figure 1 describes the process graphically using the following steps:

1. *Gathering preferences*: Each expert provides his/her preferences over the alternatives.
2. *Computing agreement level*: The consensus level is computed. Several ways to compute the consensus can be used [9].
3. *Consensus control*: The comparison between the consensus level obtained and a predefined threshold of consensus, which represents the minimal level of consensus that must be reached, is carried out. If the consensus level reach is greater or equal than the threshold, a selection process starts to select the best alternatives/s unless, another discussion round is necessary. The number of rounds also is predefined.
4. *Feedback generation*: When the minimal consensus is not reached a process to increase the level of agreement between experts begins. In such process, which can be carried out by a moderator [9], the assessments in disagreement are detected, thereafter the moderator makes suggestions to the experts for changing their preferences. Some models do not consider a moderator and the changes will be applied automatically.

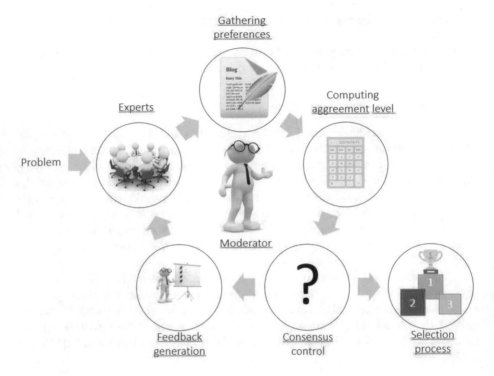

Fig. 1. CRP scheme

3   Consensus Framework

AFRYCA was developed as a Eclipse RCP application. Thus, to extend the
AFRYCA functionality, it is necessary to develop new plug-ins which are
included in the framework. This section, divided into several subsections, introduces the improvements in AFRYCA 3.0 by means of the development of
new plug-ins. Subsection 3.1 presents changes in the user interface, Subsect. 3.2
introduces the capacity to deal with multi-criteria GDM problems. Finally,
Subsect. 3.3 describes *AFRYCA Scripting Environment* (ASE).

3.1   Visualization

Analyse the CRP performance is not a easy task, since numerous aspects such
as the consensus models and its configuration have to be taken into account. For
this reason, it is necessary to provide a suitable user interface (UI) to facilitate
such analysis. Previous AFRYCA versions provided a limited UI, in which the
users could just see a set of data and results showed in plain text, without
any interaction between UI and user. The new UI, allows to the users interact
with the framework in a more intuitive way and shows interesting additional

information related to the analyses. New plug-ins for visualization have been
included in the new version, each one with its own functionality and role in the
resolution of a GDM problem. The plug-ins will be explained below in further
detail:

– *Group decision making plug-in* (see Fig. 2): this plug-in allows to visualize
the different problems already defined and saved in the application, allowing also the creation of new GDM problems. When a problem is selected,
the elements which compose the problem, experts, alternatives, criteria and
domains (if applicable), are showed. Furthermore, the experts' preferences
are also visualized, allowing to modify them at any time. Hence, to perform
a simulation, the first step is carried out in this plug-in, the GDM problem
definition.

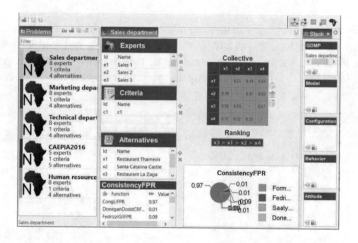

Fig. 2. Group decision making screen

– *Consensus models plug-in* (see Fig. 3): this plug-in provides the consensus
models which can be selected, classified depending on if they use a feedback
mechanism or not. When a consensus model is selected, a brief description,
main features and literature reference of the model are shown along with its
configurable parameters. The selection of the type of experts' behaviour is
also available. The next step of the simulation is performed by this plug-in,
in which the user selects the consensus model to utilize and configures the
parameters of such model to his/her liking. Afterwards, and if it is necessary,
he/she will establish and configure the experts' behaviour. Once everything
is ready, the user would initialize the simulation clicking the button located
in the upper right part of the screen.
– *Simulation plug-in* (see Fig. 4): this plug-in computes the results in the CRP.
The user can form a first judgment of the simulation, analyzing the data
which are showed, such as preferences visualization round by round through

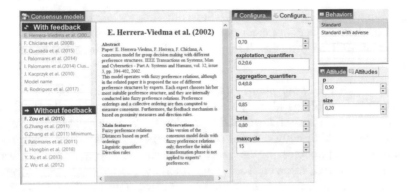

Fig. 3. Consensus models screen

MDS [5] or PCA [15] representation, alternatives ranking, solution set, metrics or plots. In addition, any simulation performed is stored and showed in this screen. When a user executes a simulation, this appears in the left part of the screen, identified by the time-stamp in which the execution has been done.

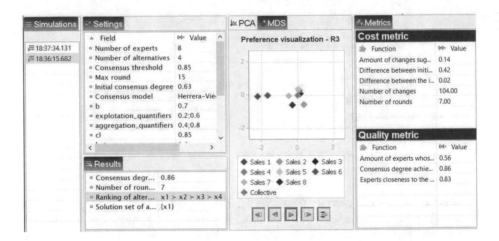

Fig. 4. Simulation screen

- *Solution plug-in* (see Fig. 5): this plug-in computes the results obtained in the CRP simulation round by round. The user can check the CRP evolution, selecting two distinct rounds and a specific expert, being able to see the evolution of such experts' preferences between the two rounds selected. To make comparisons between collective preferences is possible too.
- *AFRYCA scripting environment* (see Fig. 7): this screen allows to the users develop new features for the framework through the programming of scripts in JavaScript language. This screen will be further explained in Sect. 3.3.

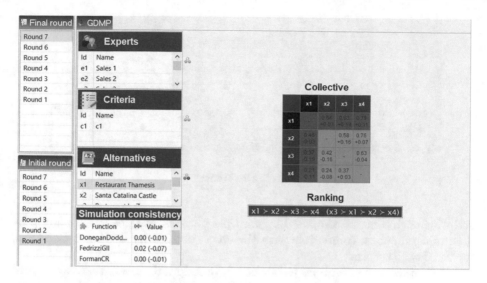

Fig. 5. Solution screen

3.2 New GDM Problems Support

Previous versions of AFRYCA managed a concrete type of GDM problems. The experts' preferences could only be provided by means of FPRs, other kind of information was not supported. Furthermore, in such problems, they were only taken into account an unique criterion impeding the processing of multi-criteria problems. To overcome these limitations, AFRYCA 3.0 is able to deal with several types of information and criteria. AFRYCA 3.0 works with FPRs [8], LPRs [10], HPRs [16], HLPRs [11,12] and DMs and, in relation to this, four different domains to express the valuation in such preference relations can be defined; fuzzy linguistic term sets (see Fig. 6), real and integer numerical and hesitant linguistic fuzzy set. Apart from supporting distinct types of information, AFRYCA 3.0 is also capable to deal with multi-criteria problems [3], being possible to define as criteria as you want. As a result of this, the amount of consensus models which can be included in AFRYCA 3.0 increases considerably.

3.3 AFRYCA Scripting Environment

The main reason to develop AFRYCA under a components architecture was to provide to the researchers a software tool easy to use and adapt to their necessities. By means of an component architecture is possible to simplify the modification and extension of the software. Taking this into account, AFRYCA 3.0 expects to enhance this concept, allowing the researcher to include new features or mathematical algorithms to the framework from the framework itself. *ASE* (AFRYCA Scripting Environment) is example of this approach. ASE is a

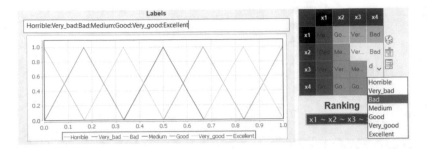

Fig. 6. Preferences

scripting environment, composed by multiple plug-ins, enables the use any programming language compatible with the Java specification JSR 223: Scripting for the Java Platform[2].

ASE utilizes the *Nashorn* engine for JavaScript. However, there are many programming languages today and researchers might not have knowledge about JavaScript and choose an alternative. For this reason, AFRYCA 3.0 provides the option of programming in multiple languages, thanks to plug-ins that it has been incorporated. Such plug-ins allow to execute source code in programming languages such as R, Groovy, Lua, Ruby, Python or Scala what gives greater

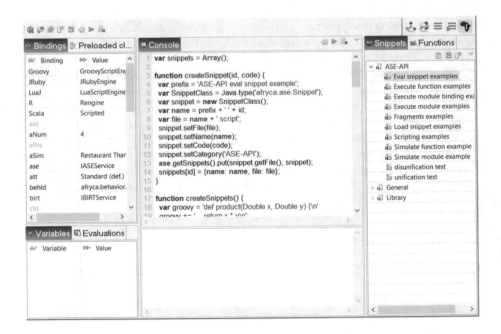

Fig. 7. ASE

---

2 https://www.jcp.org/en/jsr/detail?id=223.

freedom to users.

Through ASE, users can modify the framework by means of programming scripts in any point of the application. Researchers, developing, invoking and saving their own scripts, might increase easily the AFRYCA 3.0 functionality, incorporating, for instance, new consensus models, metrics, plots, behaviours and as consequence, personalize his/her framework.

Another strong point of ASE is that, thanks to the programming scripts, share new functionalities between different frameworks is easy, since simply incorporating the scripts files in a specific AFRYCA folder, ASE is able to recognize and incorporate them automatically to the framework.

4 Conclusions

The difficulty of the experts for reaching an agreement in real-world GDM problem makes necessary to include a CRP in its resolution. The large number of existing CRP models, each one with its specific characteristics and configurations, makes difficult the selection of the most suitable for a specific GDM problem. In this contribution a new version of the framework AFRYCA has been presented. AFRYCA 3.0 simulates CRPs for GDM problems with additional types of information and multiple criteria, also, the new user interface and the scripting environment that AFRYCA 3.0 incorporates, allow to the researchers use the framework in a more intuitive and easy way, even being able to develop new functionality from the framework itself in several programming languages, resulting a framework extremely powerful and unique.

This new version of AFRYCA gives rise to multiple future works such as, include new consensus models using different types of preference relations, new metrics to evaluate the performance of the CRPs, new kinds of visualization for preference relations and incorporate new programming languages for the scripts development in ASE.

References

1. Herrera, F., Herrera-Viedma, E., Verdegay, J.: A sequential selection process in group decision making with linguistic assessments. Inf. Sci. **85**(4), 223–239 (1995)
2. Herrera-Viedma, E., Cabrerizo, F., Kacprzyk, J., Pedrycz, W.: A review of soft consensus models in a fuzzy environment. Inf. Fusion **17**, 4–13 (2014)
3. Ishizaka, A., Nemery, P.: Multi-criteria Decision Analysis: Methods and Software. Wiley, Hoboken (2013)
4. Kacprzyk, J.: Group decision making with a fuzzy linguistic majority. Fuzzy Sets Syst. **18**(2), 105–118 (1986)
5. Kruskal, J.B., Wish, M.: Multidimensional Scaling, vol. 11. Sage Publications, New York (1978)
6. Labella, Á., Estrella, F.J., Martínez, L.: Afryca 2.0: an improved analysis framework for consensus reaching processes. Prog. Artif. Intell. **6**(2), pp. 1–14 (2017). https://doi.org/10.1007/s13748-016-0108-y

7. Lu, J., Zhang, G., Ruan, D., Wu, F.: Multi-Objective Group Decision Making. Imperial College Press, London (2006)
8. Orlovsky, S.: Decision-making with a fuzzy preference relation. Fuzzy Sets Syst. **1**(3), 155–167 (1978)
9. Palomares, I., Estrella, F., Martínez, L., Herrera, F.: Consensus under a fuzzy context: taxonomy, analysis framework AFRYCA and experimental case of study. Inf. Fusion **20**, 252–271 (2014)
10. Rodríguez, R.M., Espinilla, M., Sánchez, P.J., Martínez, L.: Using linguistic incomplete preference relations to cold start recommendations. Internet Res. **20**(3), 296–315 (2010)
11. Rodríguez, R.M., Martínez, L., Herrera, F.: A group decision making model dealing with comparative linguistic expressions based on hesitant fuzzy linguistic term sets. Inf. Sci. **241**(1), 28–42 (2013)
12. Rodríguez, R., Labella, A., Martínez, L.: An overview on fuzzy modelling of complex linguistic preferences in decision making. Int. J. Comput. Intell. Syst. **9**, 81–94 (2016)
13. Roubens, M.: Fuzzy sets and decision analysis. Fuzzy Sets Syst. **90**(2), 199–206 (1997)
14. Saint, S., Lawson, J.R.: Rules for Reaching Consensus: A Modern Approach to Decision Making. Jossey-Bass, San Francisco (1994)
15. Wold, S., Esbensen, K., Geladi, P.: Principal component analysis. Chemometr. Intell. Lab. Syst. **2**(1–3), 37–52 (1987)
16. Xia, M., Xu, Z.: Managing hesitant information in GDM problems under fuzzy and multiplicative preference relations. Int. J. Uncertainty Fuzziness Knowl. Based Syst. **21**(06), 865–897 (2013)

An Application of Transfer Learning for Maritime Vision Processing Using Machine Learning

Jeffrey W. Tweedale[✉]

Weapons and Combat Systems Division, Defence Science and Technology Group,
Edinburgh, SA 5111, Australia
Jeffrey.Tweedale@dst.defence.gov.au

Abstract. Computer Vision (CV) is a field of research that enables computers to understand their environment using digital imagery. Machine Learning (ML) processes are also used to transform data into information and Convolution Neural Network (CNN) models are currently emerging as the tool of choice for object recognition within visual images. A CNN is typically trained using supervised learning techniques. This training currently involves the manual preparation of a significant amount of data and computational effort, however the knowledge produced is portable. This paper explores the process of applying Transfer Learning (TL) to assess if it is possible to augment the pre-trained knowledge of five award winning CNNs to successfully recognise ships in the maritime environment.

Keywords: Artificial Intelligence · Convolution Neural Network
Computer Vision · Deep Learning · Machine Learning
Transfer Learning

1 Introduction

Over the last five decades, Artificial Intelligence (AI) [1,2] has expanded into many fields, such as knowledge representation, inference, Machine Learning (ML), Computer Vision (CV) and robotics. AI applications access data from the environment, and use this to reason and make decisions. Multiple systems are often combined to simultaneously provide reasoning, learning, planning, speech recognition, CV, and natural language understanding.

In the 1980's, research into ML expanded in order to investigate the performance of machines conducting nominated task(s). Topics include data representation, acquisition, categorization and classification with the aim of generating knowledge [3]. ML now includes techniques that sponsor symbolic representation, Motion Imagery (MI) and CV. These techniques enable scientists to design on-board intelligence that is essential in providing machines with the ability to recognise and understand dynamic environments [4]. ML also offers the ability to generate rich meta data that can be used by machines to learn or adapt within a given context and reduce communications bandwidths.

I. Czarnowski et al. (Eds.): KES-IDT 2018, SIST 97, pp. 87–97, 2019.
https://doi.org/10.1007/978-3-319-92028-3_9

2   Background

The concepts associated with Transfer Learning (TL) are not new. TL has been discussed by researchers in cognitive science fields since the concept was introduced by Thorndike and Woodworth early last century [5]. When applied to the ML domain, TL relates to storing knowledge gained while solving one problem and applying it to help solve a different problem with related issues [6].

For this discussion, TL is considered to be a process that is used to extend the knowledge of existing Convolution Neural Networks (CNNs) to create applications capable of predicting new labels with less effort. Huang et al. recently investigated using TL to train the reconstruction pathway or auto-encoders that are used during Synthetic Aperture Radar (SAR) target classification tasks [7]. Here the Tactical Systems Integration (TSI) Branch is assisting Navy with integration challenges for fitting Unmanned Aerial System (UAS) onto ships. Navy is considering a number of fixed and rotary wing platforms to enhance its Intelligence, Surveillance and Reconnaissance (ISR) capability. A key aspect of this integration is in the exploitation of video imagery from low attitude observations. Typically video imagery interpretations rely on human operators to engage in persistent visual target detection, classification, identification and tracking. They need to constantly monitor the observed environment. Similarly, the same operator must accumulate visual cues over time in order to achieve higher-order functions including monitoring situations or determining patterns-of-life.

Over the past decade ML has evolved as the tool of choice for object recognition. Traditional CV techniques have been chosen to enable analysts to identify objects with approximately 60% accuracy. By adopting emerging ML techniques, researchers are now approaching 97% accuracy. There are a number of parallel research streams that support ML and some authors have provided public access to their CNNs. Many of the models rely on commercial crowd-sourcing activities to generate manually pre-processed data. For example, datasets are being provided via Google, Facebook and Microsoft.

The aim of this research is to transfer trained knowledge from five award winning CNN models in order to recognise ships in the maritime environment. This paper discusses the underpinning ML processes in Sect. 3, introduces five 'State of the Art' CNN models in Sect. 4, and highlights the concepts associated with TL in Sect. 5. The TL experiments are discussed in Sect. 6, while the conclusion is provided in Sect. 7 with future comments in Sect. 8.

3   Machine Learning

As suggested above, ML involves techniques that transform data into information. This data must be collected and pre-processed into a structured dataset that can be grammatically analysed. Publicly available datasets include: the Mixed National Institute of Standards and Technology (MNIST), Canadian Institute for Advanced Research (CIFAR), Street View House Numbers (SVHN), Pattern Analysis Statistical modelling and Computational Learning (PASCAL), ImageNet and Common Object in Context (COCO). Thanks to the open source nature of the Large

Scale Visual Recognition Competition (LSVRC) researchers are able to enhance the achievements of their predecessors by fine-tuning or extending the capability provided. In this case, the major problem was finding sufficient content to re-train the number of CNN. In the CV domain this includes gathering the appropriate content to conduct training and create the most effective framework to be trained. Researchers are experimenting with Artificial Neural Networks (ANNs), CNNs and Recurrent Neural Networks (RNNs) using a combination of structures in the desire to achieve higher levels of visual recognition. The range of objects being identified continues to grow, however most current datasets provide very few maritime classification labels.

Figure 1 provides details about the winning models of the ImageNet Large Scale Visual Recognition Competition (ILSVRC) since 2010 [8]. It should be noted that the 2010 and 2011 entries used traditional CV techniques (the number of CNNs are Depicted using the red diamonds). While AlexNet produced a stepped improvement in detection performance in 2012 using five CNN layers [9]. This rate has progressively improved since 2012, with designers employing a variety of CNN architectures.

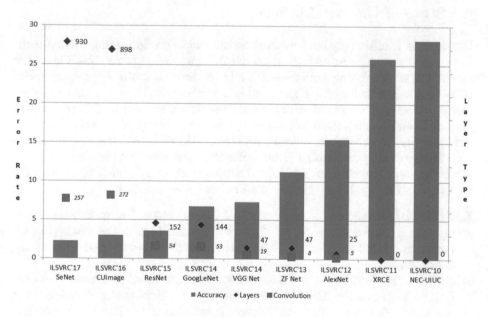

Fig. 1. ILSVRC winners (Modified)

Table 1 shows the statistics for the winning teams of the ILSVRC competition since 2012. VGG Net is mentioned as a close second in 2014, as both VGG-16 and VGG-19 Models have been used during this set of experiments. The accuracy results for VGG-16 (75.78 %) were significantly lower than VGG-19 (90.10 %).

Table 1. Winning models

Year	Model	Error rate	Accuracy	Contributor(s)	Affiliation
2012	AlexNet [9]	15.4	84.60%	Krizhevsky et al.	University of Toronto
2013	ZF Net [10]	11.2	88.80%	Zeiler and Fergus	New York University
2014	VGG Net [11]	7.3	92.70%	Simonyan and Zisserman	University of Oxford
2014	GoogLeNet [12]	6.7	93.30%	Szegedy et al.	University of Oxford
2015	ResNet [13]	3.6	96.40%	MRSA	Microsoft
2016	CUImage [14]	2.99	97.01%	Trimps-Soushen	Hong Kong
2017	SENet [15]	2.25	97.75%	Fu et al.	Momenta (Uni Oxford)

4 State of the Art Models

Researchers traditionally encode heuristic information using traditional AI techniques that produce results that approach 70% prediction accuracies. Given the recent rise of processing power available in modern Graphics Processing Units (GPUs), the results for both supervised and unsupervised learning have dramatically increased. The award winning CNN models discussed in this paper now employ Deep Learning (DL) techniques to achieve results that recently surpassed 97%. For instance, 2012 introduced AlexNet, 2014 delivered the Visual Geometry Group (VGG) and GoogLeNet, while 2015 launched ResNet. An ensemble of entries were launched in 2016 with CUImage winning, while SENet recently won the 2017 competition. Further detail about each includes:

LeNet: The first successful applications of Convolutional Networks were developed to recognise hand written digits by Yann LeCun in the 1990s [16]. Of these, the best known is the LeNet architecture that was used to read zip codes, digits.

AlexNet: The first work that popularized Convolutional Networks in Computer Vision was the AlexNet, developed by Alex Krizhevsky, Ilya Sutskever and Geoff Hinton [9]. The AlexNet was submitted to the ImageNet ILSVRC challenge in 2012 and significantly outperformed the second runner-up (top 5 error of 16% compared to runner-up with 26% error). The Network had a very similar architecture to LeNet, but was deeper, bigger, and featured Convolutional Layers stacked on top of each other (previously it was common to only have a single convolutional layer always immediately followed by a pooling layer). The winning network took between five and six days to train on two GTX 580 3GB GPU (Fermi) [9]. As shown in Fig. 2, the architecture of the AlexNet CNN contains eight learned layers, five convolution layers and three fully-connected layers. Krizhevsky et al. demonstrated that a large, deep CNN is capable of achieving record-breaking results on a highly challenging dataset using purely supervised learning. Ongoing experiments on video sequences are still being conducted in an attempt to capture helpful information that is missing or far less obvious in static images.

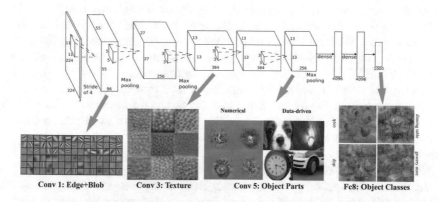

Fig. 2. Architecture of the AlexNet CNN [17]

ZF Net: The ILSVRC 2013 winner was a Convolutional Network from Matthew Zeiler and Rob Fergus [10]. It became known as the ZFNet (short for Zeiler & Fergus Net). It was an improvement on AlexNet by tweaking the architecture hyperparameters, in particular by expanding the size of the middle convolutional layers and making the stride and filter size on the first layer smaller.

GoogLeNet: The ILSVRC 2014 winner was a Convolutional Network from Szegedy et al. from Google [12]. Its main contribution was the development of an Inception Module that dramatically reduced the number of parameters in the network (4M, compared to AlexNet with 60M). Additionally, this paper uses Average Pooling instead of Fully Connected layers at the top of the ConvNet, eliminating a large number of parameters that do not seem to matter much. There are also several follow-up versions to the GoogLeNet, most recently Inception-v4.

VGGNet: The runner-up in ILSVRC 2014 was the network from Simonyan and Zisserman that became known as the VGGNet [11]. Its main contribution was in showing that the depth of the network is a critical component for good performance. Their final best network contains 16 CONV/FC layers and, appealingly, features an extremely homogeneous architecture that only performs 3x3 convolutions and 2x2 pooling from the beginning to the end.

Their pre-trained model is available for plug and play use in Caffe. A downside of the VGGNet is that it is more expensive to evaluate and uses a lot more memory and parameters (140M). Most of these parameters are in the first fully connected layer, and it has since been found that these can be removed with no performance penalty, this produces a significant reduction in the number of weighted parameters.

ResNet: Residual Network developed by Kaiming He et al. was the winner of ILSVRC 2015 [13]. It features special skip connections and a heavy use of batch normalization. The architecture is also missing fully connected layers at the end of the network. These features make the ResNet architecture a leading edge ML tool that is commonly chosen to generate ConvNets.

CUImage: Residual Network developed by Zeng et al. was the winner of ILSVRC 2016 [14]. Based on image classification models like Inception, Inception-Resnet, ResNet and Wide Residual Network (WRN), that predicts the class labels of the image. Then we refer to the framework of faster R-CNN to predict bounding boxes based on the labels. Results from multiple models are fused in different ways, using the model accuracy as weights.

SeNet: The Squeeze-and-Excitation (SE) block was produced by Hu et al. from Momenta and Oxford University was the winner of ILSVRC 2017 [15]. As the winner has just been announced, little is known about SeNet. At 930 layers, this state-of-the-art design achieves the highest prediction rates of any previously entered into the competition (although with a slight increase in computational cost). More information will be provided when they publish their research.

5   Transfer Learning

Research in CV using ML techniques is growing rapidly but the acquisition of sufficient training data is a major challenge. To create a classifier, labelled data is often used to help isolate the salient object. These classifiers are not restricted to object recognition. For instance the Caffe tutorial from Berkley provides a reasonable template on extending 'Style Recognition', by training 20 classification categories using 80,000 images (See http://caffe.berkeleyvision. org/gathered/examples/finetune_flickr_style.html). For this paper, the focus is on extending the capability of existing models to recognise ships. For example, Alexnet uses ImageNet to classify 1000 categories of object within a given scene, but only classifies generic ship labels (such as; wreck, container ship, liner, fire boat, life boat, submarine, speed boat, aircraft carrier, pirate and paddle wheel). By applying TL to AlexNet and using the existing *feature extraction* capabilities, it is possible to re-train and *fine-tune* the output layers to address new categories.

In ML, the concept of TL is the label applied to re-using existing knowledge to improve learning in an associate domain. To ensure the TL process is robust and repeatable, the same steps should be applied to all models used. These experiments are conducted in the Matlab environment to allow the novice access to the research. ZF-Net, CUImage and SeNet were not available in 'ConvNet' format for Matlab, therefore only AlexNet, GoogLeNet, VGG-16, VGG-19 and ResNet-50 were trained. In each case, the original pre-trained model (architecture and weights that recognise 1,000 object classes) was loaded into memory and the last three neural network layers removed. These were replaced and were successfully re-trained to recognise ten categories of ship. The class labels used and the number of images available included the Alvaro de Bazn (372), Anzac (428), Arleigh Burke, (1435), Commercial Ships (696), De Zeven Provincien (393), F124 Sachsen (373), Halifax (777), La Fayette (320), Santa Mara (341), and UK Type 23 (1230). It is unclear how many images are required to provide robust recognition, however figures between 5,000 (with augmentation) and 20,000 images are stated in the literature. For these experiments, feedback was provided via validation and by visualising the output layers of the re-trained CNN.

6   The Transfer Learning Experiments

This paper describes the process of applying TL to a number of world class CNN models to see if they could be used to recognise a greater number of ships in a maritime environment. Given the complex nature of each architecture, the same dataset and classification labels were used for all experiments. Similarly all variables used to extend the pre-trained knowledge were also fixed. The same hardware and software was used for all five experiments (one for each CNN model) and the results proved to be very encouraging.

The PASCAL Visual Object Classes (VOC) and Imagenet datasets were both analysed to assess if they were capable of supporting more refined classifications for ships. Unfortunately generic class labels are used. These may have initially satisfied the novice, but the enthusiasts and practitioners are now seeking more refinement. This is not an easy problem to solve, because the Lloyd's Register of Shipping (now IHS Markit) indicates that there are currently 190,000 ships above 100 gross tonnes in 21,943 categories (See www.ihs.com/products/maritime-world-ship-register.html.). This task can be decomposed by functional class, such as cargo, passenger, commercial, military or service vessels, and non-commercial or miscellaneous ships. Similarly the functional classes can be decomposed by function group and eventually by type within a class. For instance, military vessels contain fighting ships of different size, type and capability. For these experiments, research into Frigates from all nations provided sufficient scope to create a small dataset of images sourced from the internet. This was a manual process and images were primarily taken from land or other platforms, and therefore portray low viewing angle of each ship with significant background noise. A total of 6,365 images where collected and divided into ten classes (nine Frigates and one generalised background class of commercial vessels). Both PASCAL and ImageNet were pre-processed by humans. The collection of ship images were simply stored in raw format (For instance, without being manually cropped or annotated with bounding boxes). Both techniques would have helped to highlight the salient object and ensure the ship is centre normalized (within the frame) when input to the CNN. The same 'Frigate Ships' dataset was used to verify the viability of re-using pre-trained knowledge generated by each of the five state-of-the-art ML models available in 'ConvNet' format for Matlab.

The computer used consisted of a HP Z800 workstation with 12 64-bit i7 X5680 cores running at 3.33 GHz and 24 GB of physical memory running on Windows 7. The computer was fitted with a single GTX GeForce 1080 graphics card with compute 6.1 capability and a 'MaxThreadBlocksize' of [1024 1024 64] (PASCAL). Matlab 9.3.0 (2017B) was installed with a number of toolboxes and add-in packages to support CV and ML.

The approach used was to isolate 5% of the 'Frigate Ships' dataset for validation and shuffle the order within batches of 32 for the remaining images every training cycle. The learning rate for all experiments was fixed at 0.001. This was reduced 'piecewise' each epoch, while L2 regularisation was set to 90% and Momentum at $1.0\,e^{-4}$. Table 2 lists the results obtained during each experiment. The number of iterations changes with respect to the weights in the model. The number of layers also varied, however only the convolutional layers are trained as shown by the variation in network size in the last three columns. The classification results are encouraging and future experiments are planned to determine if synthetic sources can be used to enhance prediction or extend the scale. More research is required to enhance the classification accuracy.

Table 2. Experiment results

Model name	Training iterations	Model layers	Conv layers	Classification accuracy	Original network	Trimmed network	Trained network
AlexNet	1590	25	5	87.43%	222,088 kB	207,216 kB	222,088 kB
VGG-16	1273	41	16	75.78%	502,978 kB	488,074 kB	502,978 kB
VGG-19	1273	47	16	90.10%	522,360 kB	507,456 kB	522,978 kB
GoogLeNet	1297	144	53	91.35%	25,973 kB	22,237 kB	25,973 kB
ResNet	1790	177	54	91.04%	93,668 kB	86,191 kB	93,668 kB

Figure 3 shows a number of test images together with their original label and predicted classification. In most cases a 100% confidence was reported. By visually examining the results, the class with the least number of training images indicated the largest number of classification errors. More images are being sought to verify this observation.

Fig. 3. Classification results following output training using Transfer Learning

7   Conclusion

This paper highlights the successful use of TL when training a CNN to recognise military frigates (ships) in the maritime domain. Existing research indicates that millions of labelled examples are needed to successfully recognise objects in visual scenes. However these experiments show that encouraging classification results can be achieved using TL. The experiments also show that the architecture influences the accuracy of the model. The results also show that there were insufficient ship images to fully re-train the models. That is, six classes contained less than 500 images. These errors are highlighted when classifying the 'Santa Maria' class frigate. If this class is removed the classification accuracy rises from 91% to 96.86%.

Confidence in classification accuracy also increases when the final layer is visualised. Many of the trained outputs in the original 1000 imageNet classes are well defined (especially those with good contrast). Unfortunately it's difficult to see this level of definition in the 'frigates ships' dataset.

Although there is room for improvement, these experiments demonstrate that it is possible to transfer existing knowledge from CNN models and achieve similar results. For instance the classification accuracies for these experiments varied between 2.83% (84.6% minimum) and −5.0% (91.04% maximum). This training verifies the original models and validates the ability to apply TL to recognise ships in the maritime domain.

8   Future and Recommendations

Future experiments will explore growing the dataset by augmenting the repository with modified versions of current images. This includes modifying the intensity, colour, cropping, shifting, flipping and rotating the original image. Other techniques can be used, however this author has begun investigating the ability to synthesise images using 3D computer generated renderings and taking pictures using calibrated camera runs from a variety of range and elevation settings in fixed steps around its centre. Until then, more information is sought by networking through forums or by attending ILSVRC and other leading edge conferences. As the community grows, new techniques will emerge to enable researchers to pursue suitable ML architectures capable of conducting situation assessment and ultimately generating Tactical Knowledge.

References

1. Grevier, D.: AI - The Tumultuous History of the Search for Artificial Intelligence. Basic Books, New York (1993)
2. Callan, R.: Artificial Intelligence. Palgrave MacMillan, Hampshire (2003)
3. Tambe, M., Pynadath, D., Chauvat, C., Das, C., Kaminka, G.: Adaptive agent architectures for heterogeneous team members. In: International Conference on Multi-Agents Systems (ICMAS2000), Boston, MA, July 2000
4. Galway, L., Charles, D., Black, M.: Machine learning in digital games: a survey. Artif. Intell. Rev. **29**, 123–161 (2008)
5. Thorndike, E.L., Woodworth, R.S.: The influence of improvement in one mental function upon the efficiency of other functions. Psychol. Rev. **8**(3), 247–261 (1901)
6. Dietterich, T.: Special issue on inductive transfer. J. Mach. Learn. **28**(1) (1997)
7. Huang, Z., Pan, Z., Lei, B.: Transfer learning with deep convolutional neural network for sar target classification with limited labeled data. Remote Sens. **9**(907), 1–21 (2017)
8. Russakovsky, O., Deng, J., Su, H., Krause, J., Satheesh, S., Ma, S., Huang, Z., Karpathy, A., Khosla, A., Bernstein, M., Berg, A.C., Fei-Fei, L.: ImageNet large scale visual recognition challenge. Int. J. Comput. Vis. (IJCV) **115**(3), 211–252 (2015)
9. Krizhevsky, A., Sutskever, I., Hinton, G.E.: ImageNet classification with deep convolutional neural networks. In: Proceedings of the 25th International Conference on Neural Information Processing Systems, NIPS 2012, USA. pp. 1097–1105. Curran Associates (2012)
10. Zeiler, M.D., Fergus, R.: Visualizing and understanding convolutional networks. CoRR abs/1311.2901 (2013)
11. Simonyan, K., Zisserman, A.: Very deep convolutional networks for large-scale image recognition. CoRR 1409.1556, pp. 1–14 (2014)
12. Szegedy, C., Liu, W., Jia, Y., Sermanet, P., Reed, S.E., Anguelov, D., Erhan, D., Vanhoucke, V., Rabinovich, A.: Going deeper with convolutions. CoRR abs/1409.4842 (2014)
13. He, K., Zhang, X., Ren, S., Sun, J.: Deep residual learning for image recognition. CoRR abs/1512.03385 (2015)

14. Zeng, X., Ouyang, W., Yan, J., Li, H., Xiao, T., Wang, K., Liu, Y., Zhou, Y., Yang, B., Wang, Z., Zhou, H., Wang, X.: Crafting GBD-net for object detection. IEEE Trans. Patt. Anal. Mach. Intell. **99**, 1 (2017)
15. Hu, J., Shen, L., Sun, G.: Squeeze-and-excitation networks. CoRR abs/1709.01507 (2017)
16. LeCun, Y., Bottou, L., Bengio, Y., Haffner, P.: Gradient-based learning applied to document recognition. Proc. IEEE **86**, 2278–2324 (1998)
17. Wei, D., Zhou, B., Torralba, A., Freeman, W.T.: mNeuron: A Matlab Plugin to Visualize Neurons from Deep Models. Massachusetts Institute of Technology (2017)

Fuzzy Regression Model Dealing with Vague Possibility Grades and Its Characteristics

Yoshiyuki Yabuuchi$^{(\boxtimes)}$

Shimonoseki City University, 2-1-1 Daigaku-cho, Shimonoseki,
Yamaguchi 751-8510, Japan
`yabuuchi@shimonoseki-cu.ac.jp`

Abstract. Given that an interval-type fuzzy regression model illustrates the possibilities of an analysis target according to its intervals, its characteristics can be intuitively understood. Conversely, there is the important problem of properly describing the possibilities of an analysis target. In other words, a fuzzy regression model should be designed to illustrate an appropriate possibility interval according to data. We continue to propose models and approaches that illustrate the appropriate possibility intervals. These are models that correspond to samples that distort possibility intervals, maximize the sum of possibility grades obtained from an interval-type fuzzy regression model and so on. Thanks to various improvements, we obtained a model illustrating the appropriate possibility intervals. On the other hand, by using the possibilities of unusual samples, the centers of the model and the data distribution do not coincide and the possibility intervals might be distorted. For this reason, we assumed vagueness was included in the possibility grades as well as the proposed fuzzy regression model dealing with that vagueness. The proposed model can be obtained only by deciding the extent to which sample possibilities are considered. By verifying the model using a numerical example, the features were found. The appropriate possibility interval can be obtained by setting restrictions on the number of samples that are neglected during model construction. Then, by moderately increasing the magnitude of vagueness included in the possibility grades, we can manage distorted possibility intervals. This paper discusses the results obtained.

Keywords: Interval-type fuzzy regression · Vagueness
Possibility grade

1 Introduction

Computer networks are used all over the world, and enormous amounts of various data are continuously flowing on the network. Although it is called big data, there is a movement to try to acquire the best big data processing by utilizing

© Springer International Publishing AG, part of Springer Nature 2019
I. Czarnowski et al. (Eds.): KES-IDT 2018, SIST 97, pp. 98–108, 2019.
https://doi.org/10.1007/978-3-319-92028-3_10

this in the business fields. However, utilization of big data has been delayed for reasons such as a lack of data scientists and introduction costs. In fact, there are reports that 52.1% of companies in Japan do not have a data scientist [4]. Therefore, it cannot be said that the information obtained from a statistical model is employed.

A regression model is a statistical tool used in various fields such as management and economics. Errors contained in observed values can be processed statistically. If it were the case that a person would be included as a part of the elements to be analyzed, a soft computing method based on fuzzy theory would be effective for human vagueness [7,8,10]. Soft computing has a fuzzy regression model, which can be roughly divided into interval and non-interval type fuzzy regression models. An interval-type fuzzy regression model describes the intervals in which the dependent variable can be observed in its interval [5,9]. This provides a model by including all samples. However, it is also difficult to analyze in detail the relationship between variables and vagueness included in samples. However, the intervals of an interval model make it possible to intuitively determine the characteristics of the analysis target. Conversely, a non-interval model shows the relationships between independent and dependent variables by considering the vagueness of an analyzed system [1–3,6]. In addition, non-interval models can analyze the relationships between variables and vagueness included in samples like statistical models.

As mentioned above, in business and economy, it can be thought that there are many opportunities to use an interval-type fuzzy regression model that can intuitively understand the characteristics of an analysis target. However, if the amount of vagueness included in samples is large, an interval-type fuzzy regression model blurs the possibility of an analyzed system indicated by intervals, and its shape is greatly distorted. We proposed a Fuzzy Regression Model Dealing with Vague Possibility Grades (FRVG) which properly handles the vagueness contained in samples and appropriately illustrates the possibilities of an analyzed system [15]. This model aims to minimize the influence of the shape of the data distribution and intuitively understand features of an analysis target. This paper considers the features of this model using a numerical example.

The remainder of this paper is as follows: Sect. 2 explains an interval-type fuzzy regression model and a possibility grade. Section 3 explains an interval-type fuzzy regression model using a possibility grade. Section 4 discusses the features of a fuzzy regression model dealing with a vagueness possibility grade using a numerical example and confirms its characteristics. Section 5 concludes this paper.

2   Interval-Type Fuzzy Regression Model

First, consider n sets of samples $(\boldsymbol{x}_i, y_i)$, $i = 1, 2, \ldots, n$ consisting of independent variables $\boldsymbol{x}_i = [x_{i1}, x_{i2}, \ldots, x_{ip}]$ with p attribute values and a dependent variable y_i. An interval-type fuzzy regression model with symmetric triangular fuzzy numbers $\boldsymbol{A}_j$, $j = 1, 2, \ldots, p$ as regression coefficients is as follows:

$$\boldsymbol{Y}_i = \boldsymbol{A}_0 + \boldsymbol{A}_1 X_1 + \boldsymbol{A}_2 X_2 + \cdots + \boldsymbol{A}_p X_p. \tag{1}$$

Here, fuzzy regression coefficients $\boldsymbol{A}_j = (a_j, c_j)$, $j = 1, 2, \ldots, p$, consist of a center a_j of a triangle and the width c_j. Then, the output of Eq. (1), $\boldsymbol{Y}_i$, becomes the symmetric triangular fuzzy number $\boldsymbol{Y}_i = (Y_i^C, Y_i^W)$, similar to the fuzzy regression coefficient by the extension principle. Here, since Y_i^W is a width, it is a positive value and $Y_i^W = \boldsymbol{c}|\boldsymbol{x}_i|$. The upper limit Y_i^U, the lower limit Y_i^L and the center Y_i^C of $\boldsymbol{Y}_i$ are as follows:

$$Y_i^U = \boldsymbol{a}\boldsymbol{x}_i + \boldsymbol{c}|\boldsymbol{x}_i|, Y_i^C = \boldsymbol{a}\boldsymbol{x}_i, Y_i^L = \boldsymbol{a}\boldsymbol{x}_i - \boldsymbol{c}|\boldsymbol{x}_i|.$$

Given that an interval-type fuzzy regression model describes the possibilities of an analysis target by including samples, then an interval-type fuzzy regression model is based on the vagueness of this model. That is, this can be written in LP to minimize model widths $\sum_{j=1}^p c_j$ or forecasted value widths $\sum_{i=1}^n \boldsymbol{c}|\boldsymbol{x}_i|$ as follows:

$$\begin{aligned} &\text{min. } F \\ &\text{s.t. } \quad y_i \in \boldsymbol{Y}_i, i = 1, 2, \ldots, n, \\ &\qquad c_j \geq 0, j = 1, 2, \ldots, p. \end{aligned} \tag{2}$$

As mentioned above, the objective function F of LP should be selected, such as $F_1 = \sum_{i=1}^p c_j$ and $F_2 = \sum_{i=1}^n \boldsymbol{c}|\boldsymbol{x}_i|$, according to an analysis purpose.

3   Fuzzy Regression Model Based on Possibility Grade

3.1   Fuzzy Regression Model Through Maximizing Possibility Grades

As shown in Eq. (1) and LP (2), a model center Y_i^C is a parameter for obtaining a fuzzy regression coefficient with the least vagueness. In this case, Y_i^C does not coincide with the center of data distribution. In addition, when sample y_i coincides with the center Y_i^C of possibility intervals, its possibility grade is 1. That is, an interval-type fuzzy regression model describes the possibility intervals of an analysis target, but its center does not coincide with the center of its possibility distribution. Yabuuchi and Watada proposed a fuzzy regression model that maximized possibility grades (FRMG) in order to solve this problem [11–14]. FRMG is a fuzzy regression model that maximizes the sum of possibility grades obtained from models and samples. In addition, by considering the bias of the data distribution, a fuzzy regression coefficient uses an asymmetric LR fuzzy number. For the fuzzy regression coefficient $\boldsymbol{A} = (a, c, d)$, the center denoted by a, and widths of upper and lower sides are c and d, respectively. As a result, both the limits of the upper W_i^U and lower W_i^L are as follows:

$$W_i^U = \boldsymbol{c}|\boldsymbol{x}_i|, W_i^L = \boldsymbol{d}|\boldsymbol{x}_i|.$$

Both the limits of the upper Y_i^U and lower Y_i^L of the forecasted value $\boldsymbol{Y}_i$ are as follows:

$$Y_i^U = Y_i^C + W_i^U, Y_i^L = Y_i^C - W_i^L.$$

Given that possibility grades are large, as described above, when the widths of predicted values are large, it is necessary to minimize the model width. In addition, unlike a conventional fuzzy regression model, vagueness is not the primary objective, so the model shape tends to be distorted due to the shape of the data distribution. For this reason, we did not constrain the inclusion relationship between samples and the regression model, and formulated our model using parameters as follows:

$$\text{max. } \alpha \sum_{i=1}^{n} \mu(\boldsymbol{x}_i, y_i) - (1 - \alpha)F \tag{3}$$
$$\text{s.t.} \quad c_j, d_j \geq 0, j = 1, 2, \ldots, p.$$

Normally, data distribution coincides with a model center in Eq. (3). However, depending on the shape of a data distribution, a model may not correctly describe a possibility distribution. In this case, an appropriate model can be obtained by adding a constraint expression that some samples are not included in the model. This is also effective when outliers are mixed in. Moreover, it is an effective means to further use the following membership function to adjust the influences of unusual samples.

$$\mu(\boldsymbol{x}_i, y_i) = \begin{cases} \max(0, \delta - (y_i - Y_i^C)/W_i^U), \ y_i \geq Y_i^C, \\ \max(0, \delta - (Y_i^C - y_i)/W_i^L), \ Y_i^C > y_i. \end{cases} \tag{4}$$

When constructing a model, Eq. (4) sets the range of the membership function to $[0, \delta]$ and returns the range to $[0, 1]$ after obtaining its coefficient. The data distribution is relatively small, and the influence of unusual samples is reduced.

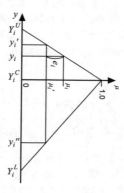

Fig. 1. Possibility grade of y_i including e_i in an LR fuzzy number

3.2   Fuzzy Regression Model Dealing with Vague Possibility Grade

Even if a fuzzy regression model is obtained using the possibility grade as described above, the influence of unusual samples cannot be removed. Given that errors are included in the observed values, it is natural to think that the possibility grade includes vagueness. In other words, a fuzzy regression model can reflect only the possibilities of samples during model construction. Figure 1 shows the relationship between the observed value y_i including the vagueness e_i and the possibility grade μ_i. When the possibility grade μ_i of y_i contains vagueness e_i, the possibility grade of removing vagueness is $\mu_i' = \mu_i - e_i$. When the vagueness e_i is a positive value, it is highly valued compared to the original grade μ_i. Therefore, when vagueness is removed from observation value y_i, a value distant from the predicted value center Y_i^C is the original value according to its vagueness. Conversely, when the vagueness e_i is negative and is evaluated to be smaller than the original grade, the observed value y_i is close to the center of the model and should be evaluated more greatly. In other words, when vagueness is a negative value and an observed value is an unusual sample, it is possible to construct a model by moving in the center direction of the data distribution. Let us consider y_i^* that removes the vagueness from y_i. First, let $L_i = Y_i^U - Y_i^C$ and $R_i = Y_i^C - Y_i^L$ be the upper and lower limit side widths in an asymmetric triangular fuzzy number. As shown in Fig. 1, y_i^*, obtained by removing vagueness e_i from y_i has two types, y_i' and y_i''. They are as follows:

$$y_i^* = \begin{cases} \{y_i + e_i L_i, y_i - (1 - \mu_i)L_i - (1 - \mu_i + e)R_i\}, & Y_i^C \leq y_i \\ \{y_i + (1 - \mu_i)R_i + (1 - \mu_i + e_i)L_i, y_i - e_i L_i\}, & y_i < Y_i^C. \end{cases} \tag{5}$$

Here, if it is a symmetric triangular fuzzy number, then both widths are $Y_i^W = L_i = R_i$.

However, $y_i - (1 - \mu_i)L_i - (1 - \mu_i + e_i)R_i$ and $y_i + (1 - \mu_i)R_i + (1 - \mu_i + e_i)L_i$ are on the opposite side of Y_i^C, therefore, they will be manipulated to a large extent. In this case, it is preferable to use $y_i' = y_i + e_i Y_i^W$ and $y_i - e_i Y_i^W$ as observations with the vagueness removed. Using these, the LP (3) can be rewritten, and we will use the improved fuzzy regression model in a numerical example. For the FRMG defined by Eq. (5), an inclusion relationship between FRMG and the samples is not included in the constraints, however, so in order to replace y_i and y_i^*, it is rewritten as follows:

$$\text{max. } \alpha \sum_{i=1}^{n} \mu(\boldsymbol{x}_i, y_i^*) - (1 - \alpha)F \tag{6}$$
$$\text{s.t. } c_j, d_j \geq 0, j = 1, 2, \ldots, p.$$

Here, since the possibility grade takes the value of $[0, 1]$, vagueness e_i should be limited to accepting a value within the range of $|e_i| \leq 1$. However, even if $|e_i|$ is greater than 1, since fuzzy regression coefficients are obtained using y_i^*, the possibility grade is the value of $[0, 1]$. Therefore, when setting the constraint condition, $|e_i| \leq \varepsilon$, pay attention to how the parameter is set.

4   Numerical Example

Here, we confirm the characteristics of the model using a numerical example generated by a random number. In order to grasp the features of the model, we use a numerical example of samples obtained by adding errors to random variables at $y = x$. In addition, one sample was changed an outlier.

The regression equation to be obtained is $Y = A_0 + A_1 X$ and the regression coefficients are $A_0 = (a_0, c_0)$ and $A_1 = (a_1, c_1)$, respectively. In Sect. 2, two representative objective functions F_1 and F_2 of a conventional fuzzy regression model were shown. In Sect. 3, we presented the improved fuzzy regression models. In this section, by comparing these, we confirm the effectiveness of the vagueness processing included in an improved fuzzy regression model or possibility grade. Here, when converting (5) unlimitedly, the possibility of results completely different from the analysis target will be described. For this reason, in this numerical example, more than 90% of samples are kept within the possibility intervals of a fuzzy regression model. In this comparison, we compare the conventional fuzzy regression model (1) and the improved fuzzy regression model for each objective function of the LPs. We also compare FRMG and FRVG. The sum of the possibility grades from the model and samples, the sum of the possibility grades for the widths of the forecasted values, and the sum of the widths of the forecasted values are used as indexes to indicate the features of the models.

Table 1. Regression coefficients and evaluation indices of model using function F_1

		$\varepsilon = 0$	$\varepsilon = 0.3$	$\varepsilon = 0.6$	$\varepsilon = 1.0$
Regression	A_0	(4.232, 0)	(4.232, 0)	(4.232, 0)	(4.495, 0)
coefficients	A_1	(0.908, 0.685)	(0.908, 0.527)	(0.908, 0.428)	(0.701, 0.435)
Index 1		21.473	19.549	17.830	16.420
Index 2		2.542	2.921	3.244	3.077
Index 3		689.042	530.032	430.651	437.480

Index 1: Sum of possibility grades derived from the model and samples
Index 2: Sum of possibility grades for widths of forecasted values
Index 3: Sum of widths of forecasted values

4.1 Conventional and Improved Models Using Function F_1

Here, $F_1 = \sum_{j=1}^{p} c_j$ is used for the objective function of LP in an interval-type fuzzy regression model. The obtained fuzzy regression coefficients and evaluation indices are shown in Table 1.

In Table 1, ε is the parameter of the constraint regarding the magnitudes of vagueness included in possibility grades, and $\varepsilon = 0$ refers to a conventional fuzzy regression model. In a conventional model, the slope of the model center is small with respect to the center of the data distribution, and they do not coincide. However, for the parameters from 0 to 0.6, the center of the coefficient is the same; however, the width of the coefficients of the independent variable decreases as the parameter increases. The data distribution and the center of the model are close as well. This is shown in Fig. 2. The broken line at the center shows the center of the model, and it appears to be consistent in the figure. In addition, the outermost dashed-dotted line is the conventional model with $\varepsilon = 0$, the width decreases as parameters increase, and any outlier will reside outside of the possibility interval.

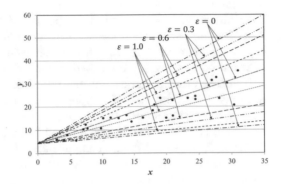

Fig. 2. Model using objective function F_1 and parameter

When the width of the membership function is large, since samples are relatively near the center of the membership function, the possibility grade is a large value. Moreover, the ratio of the width of the predicted value and the possibility grade becomes a large value as the width decreases to some extent. This can be seen from indices 2 and 3. From the above, for this model, we can control the influences of an outlier by increasing ε.

Table 2. Regression coefficients and evaluation indices of model using function F_2

		$\varepsilon = 0$	$\varepsilon = 0.3$	$\varepsilon = 0.6$	$\varepsilon = 1.0$
Regression	A_0	(8.627, 6.717)	(3.489, 1.141)	(3.340, 2.649)	(4.039, 1.244)
coefficients	A_1	(0.634, 0.016)	(1.047, 0.512)	(0.944, 0.228)	(0.859, 0.270)
Index 1		15.927	18.486	17.469	17.003
Index 2		2.287	2.385	2.965	3.569
Index 3		432.172	585.565	393.700	348.199

4.2 Conventional and Improved Models Using Function F_2

We investigate the model using $F_2 = \sum_{i=1}^{n} c|x_i|$ as the objective function of the LP. The obtained fuzzy regression coefficients and evaluation indices are shown in Table 2. In the case of using the objective function F_2, the independent variable becomes a weight, so that the width of the possibility interval becomes small where the independent variable has a large value. Conversely, the objective function F_1 has no weight, so the possibility interval around the origin becomes

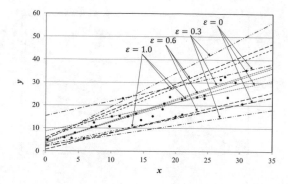

Fig. 3. Model using objective function F_2 and parameter

small. Then, the shape of the model is distorted due to the influence of an outlier for $\varepsilon = 0$ which is of a conventional type (see Fig. 3). By increasing the value of the parameter, it shifts to a value closer to the slope of the data distribution, while the width of the model is small. As a result, the width of the predicted values decreases as the parameter value increases. The possibility grade is small in accordance with the width of the predicted value in Table 2, but the sum of the possibility grade with respect to the width of the predicted value is large.

Table 3. Regression coefficients and evaluation indices of FRVG

		$\varepsilon = 0$	$\varepsilon = 0.3$	$\varepsilon = 0.6$	$\varepsilon = 1.0$
Regression	A_0	(2.659, 0.749)	(3.504, 1.357)	(3.212, 1.269)	(2.944, 0.643)
coefficients	A_1	(0.950, 0.332)	(0.954, 0.171)	(0.953, 0.155)	(0.973, 0.214)
Index 1		17.081	13.987	13.226	14.018
Index 2		3.435	3.875	3.947	4.244
Index 3		380.163	256.164	234.792	254.751

4.3   Fuzzy Regression Model Dealing with Vague Possibility Grade

At this point, we verify the relationship between the FRVG defined by Eq. (6) and the parameters. FRVG is FRMG with $\varepsilon = 0$ the same as a conventional model. Table 3 shows the obtained fuzzy regression coefficients and evaluation indices. Given that FRMG and FRVG also describe the center of a data distribution, then essentially fuzzy regression coefficients are based on LR fuzzy numbers. Here, the fuzzy regression coefficients of LL fuzzy numbers are leveraged such that they can be compared with other models. Given that FRMG can manage the influence of unusual samples with the parameters α and δ, even if $\varepsilon = 0$,

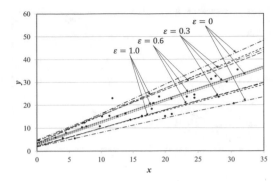

Fig. 4. FRVG and parameter

the model is not greatly distorted. However, as the value of the parameter is increased, the center of the constant term is small and the center of the gradient is large (see Fig. 4). Then, when $\varepsilon = 1.0$ was used, similar coefficients that used $\varepsilon = 0$ were obtained, however the vagueness of the model became small.

4.4   Setting the Parameter to 1.0 or More

In this case, the behavior when parameter ε is set to 1 or more is confirmed. The fuzzy regression coefficients and the evaluation indices of each model are shown in Table 4. Given that the value of the possibility grade is $[0, 1]$, it is better that the parameter ε is smaller than $[0, 1]$. However, even if the parameter is set to a large value, the shape does not distort, it becomes stable, and the vagueness of the model decreases. In addition, even if the parameter was set to a certain large value, there was no change in the model. For FRVG, the same model as $\varepsilon = 1.0$ was obtained even with $\varepsilon \geq 1.0$.

Table 4. Regression coefficients, evaluation indices, and parameters of the models

		$\varepsilon = 1.5$	$\varepsilon = 2.0$
The model	A_0, A_1	(4.232, 0), (0.908, 0.274)	(4.363, 0), (0.891, 0.230)
using F_1	Index 1, 2, 3	13.721, 3.731, 275.617	12.253, 3.912, 231.563
The model	A_0, A_1	(3.990, 2.169), (0.808, 0.138)	(4.132, 1.495), (0.786, 0.191)
using F_2	Index 1, 2, 3	13.359, 3.326, 273.368	13.483, 3.393, 284.759
The FRVG	A_0, A_1	(2.944, 0.643), (0.973, 0.214)	(2.944, 0.643), (0.973, 0.214)
	Index 1, 2, 3	14.018, 4.244, 254.751	14.018, 4.244, 254.751

5   Conclusion

An interval-type fuzzy regression model can intuitively grasp the characteristics of systems. However, depending on the shapes of the data distribution, the system characteristics may be blurred. Therefore, in order to solve this problem, we considered that the vagueness is included in the possibility grades and proposed an approach to deal with the vagueness. In this paper, we discussed the features of the proposed method. The following bullet points summarizes our discussion thus far.

- The vagueness included in possibility grades should be 1.0 or less. On the other hand, even if we set inappropriate values, the model did not distort.
- By increasing the allowable vagueness, the vagueness included in the possibility grades gets smaller, and it is possible to reduce the influence of unusual sample.
- If many samples contain vagueness, the model shape is greatly distorted.

Finally, the usefulness of the proposed model was confirmed in the application results to real data that this paper did not show. The proposed model should be further applied to actual data, and it must show its usefulness.

References

1. Coppi, R., D'Urso, P., Giordani, P., Santoro, A.: Least squares estimation of a linear regression model with LR Fuzzy response. Comput. Stat. Data Anal. **51**(1), 267–286 (2006)
2. Diamond, P.: Fuzzy least squares. Inf. Sci. **46**(3), 141–157 (1988)
3. D'Urso, P., Gastaldi, T.: A least-squares approach to fuzzy linear regression analysis. Comput. Stat. Data Anal. **34**(4), 427–440 (2000)
4. Japanese Ministry of Internal Affairs and Communications: 2017 White Paper Information and Communication in Japan (2017)
5. Lee, H., Tanaka, H.: Upper and lower approximation models in interval regression using regression quantile techniques. Eur. J. Oper. Res. **116**(3), 653–666 (1999)
6. Modarres, M., Nasrabadi, E., Nasrabadi, M.M.: Fuzzy linear regression model with least square errors. Appl. Math. Comput. **163**(2), 977–989 (2005)
7. Ramli, A.A., Watada, J., Pedrycz, W.: Real-time fuzzy regression analysis: a convex hull approach. Eur. J. Oper. Res. **210**(3), 606–617 (2011)
8. Ramli, A.A., Watada, J., Pedrycz, W.: A combination of genetic algorithm-based fuzzy C-means with a convex hull-based regression for real-time fuzzy switching regression analysis: application to industrial intelligent data analysis. IEEJ Trans. Electr. Electron. Eng. **9**(1), 71–82 (2014)
9. Tanaka, H., Watada, J.: Possibilistic linear systems and their application to the linear regression model. Fuzzy Sets Syst. **27**(3), 275–289 (1988)
10. Watada, J., Pedrycz, W.: A fuzzy regression approach to acquisition of linguistic rules. In: Pedrycz, W., Skowron, A., Kreinovich, V. (eds.) Handbook of Granular Computing, pp. 719–740. Wiley (2008)
11. Yabuuchi, Y., Watada, J.: Fuzzy regression model building through possibility maximization and its application. Innov. Comput. Inf. Control Express Lett. ICI-CEL **4**(2), 505–510 (2010)

12. Yabuuchi, Y., Watada, J.: Fuzzy robust regression model by possibility maximization. J. Adv. Comput. Intell. Intell. Informat. **15**(4), 479–484 (2011)
13. Yabuuchi, Y.: Japanese economic analysis by a fuzzy regression model building through possibility maximization. In: Proceedings of the 6th Conference on Soft Computing and Intelligence System and the 13th International Symposium on Advanced Intelligent Systems, pp. 1772–1777 (2012)
14. Yabuuchi, Y., Watada, J.: Fuzzy robust regression model building through possibility maximization and analysis of Japanese major rivers. ICICEL **9**(4), 1033–1041 (2015)
15. Yabuuchi, Y.: Possibility grades with vagueness in fuzzy regression models. In: Proceeding of KES 2017, pp. 1470–1478 (2017)

Decision-Oriented Composition Architecture for Digital Transformation

Alfred Zimmermann[1(✉)], Rainer Schmidt[2], Kurt Sandkuhl[3],
Dierk Jugel[1,3], Justus Bogner[1,4], and Michael Möhring[2]

[1] Herman Hollerith Center, Reutlingen University, Danziger Str. 6,
71034 Böblingen, Germany
{alfred.zimmermann,
dierk.jugel}@reutlingen-university.de
[2] Munich University of Applied Sciences, Lothstrasse 64,
80335 Munich, Germany
{rainer.schmidt,michael.moehring}@hm.edu
[3] University of Rostock, Albert Einstein Str. 22, 18059 Rostock, Germany
{kurt.sandkuhl,dierk.jugel}@uni-rostock.de
[4] DXC Technology, Herrenberger Str. 140, 71034 Böblingen, Germany
justus.bogner@dxc.com

Abstract. Enterprises are presently transforming their strategy, culture, processes, and their information systems to become more digital. The digital transformation deeply disrupts existing enterprises and economies. Digitization fosters the development of IT systems with many rather small and distributed structures, like Internet of Things or mobile systems. Since years a lot of new business opportunities appeared using the potential of the Internet and related digital technologies, like Internet of Things, services computing, cloud computing, big data with analytics, mobile systems, collaboration networks, and cyber physical systems. This has a strong impact for architecting digital services and products. The change from a closed-world modeling perspective to more flexible open-world composition and evolution of system architectures defines the moving context for adaptable systems, which are essential to enable the digital transformation. In this paper, we are focusing on a decision-oriented architectural composition approach to support the transformation for digital services and products.

Keywords: Digital services and products · Digital transformation
Digitization architecture · Architectural composition and evolution
Decision management

1 Introduction

Nowadays, more and more smart connected products and services are available and extending physical components capabilities using the Internet [1]. Furthermore, digitization, as introduced by Schmidt et al. [2], enables new data-driven processes and increases better decision making. Intelligent cars and smart devices are for instance part of a new digital economy with digital products, services, and processes [3].

© Springer International Publishing AG, part of Springer Nature 2019
I. Czarnowski et al. (Eds.): KES-IDT 2018, SIST 97, pp. 109–119, 2019.
https://doi.org/10.1007/978-3-319-92028-3_11

Digitization [2] requires the appropriate alignment of digital business models with digital technologies, which are synchronously directed by new digital strategies. Current digitized applications are integrating Internet of Things, Web services, REST services, Microservices, cloud computing, big data, machine learning with new frameworks and methods, emphasizing openly defined service-oriented software architectures [4] with extensions for semantic services.

Both business and technology are impacted from the digital transformation by complex relationships between architectural elements, which directly affect the adaptable digitization architecture [3] for digital products and services and their related digital governance. Enterprise Architecture Management (EAM) [5] organize, build and utilize distributed capabilities for the digital transformation [6–8].

Furthermore, the digitization process [3] demands flexibility to adapt to rapidly changing business requirements and newly emerging business opportunities. Therefore, many enterprises using concepts like Internet of Things [9, 10] and Microservice Architectures (MSA) [11, 12] to handle this fast digitization process. Applications built this way consist of several fine-grained services that are independently scalable and deployable. Using Microservice Architectures [11] organizations can increase agility and flexibility for business and IT systems, which fits better with small-sized integrated systems in the age of digital transformation.

Unfortunately, the current state of art and practice of enterprise architecture lacks an integral understanding of fast and flexible adaptation of architectures and decision management when integrating by composition of micro-granular systems and services, like Microservices and Internet of Things for the context of digital transformation and evolution of architectures. Our goal is to extend previous approaches of static closed-world enterprise architecture modeling to fit for the flexible and adaptive digitization when integrating new services and products coming from the open-world. This fundamental change from the closed to the open-world architectural modeling should be achieved by introducing suitable mechanisms for decision-oriented collaborative architectural engineering and methods for integrating micro-granular architectures by a new composition approach.

Our current paper extends our previous work and is part of an ongoing research project including different current and past research results. We investigate the following main research question in this paper:

What are suitable decision-oriented architectural composition and evolution approaches for flexible integrating and managing a huge amount of micro-granular structures, like Internet of Things and Microservices to support their flexible open-world integration as part of digital services and products?

The following Sect. 2 sets the base for the digital transformation with the context of digitized services and products. Section 3 focusses on architecting digital structures, systems, and technologies, while Sect. 4 presents suitable service-based architectural composition and software evolution approaches. In Sect. 5 we are investigating concepts and mechanisms for architectural decision management of multi-perspective digital architectures. Finally, we conclude in Sect. 6 our research findings and limitations and sketch our future research steps.

2 Digital Services and Products

Digital services [2] and associated digital products are software-intensive as well as malleable and usually service-oriented [13]. These services are able to increase their capabilities as well as change their behaviour [13]. Furthermore, digitized products can be able to support as well as increase the co-creation of the (service) value together with customers and (other) stakeholders [2].

In general, classical industrial products are static [6]. However, digitization provides the means to enrich them by digital services to be more flexible [8]. For instance, every industrial product can be complemented by services. These services can be updated and extended.

The Internet of Things [9, 10] drives the creation of digital products and services. The devices connected to the internet can exchange easily different information to support business processes. Furthermore, it is possible to get maintenance relevant data to run a better predictive maintenance approach. Therefore, more customer-oriented products will be available on the market. Furthermore, through linking data from different sources [14], it is possible to get better basements for decisions.

Additionally, platforms [15] and standardized interfaces [16] are important to support digitized services and products. Otherwise, the heterogeneity will destroy the value and the community effects of this new digitized services and products.

3 Digital Enterprise Architecture

Architecture Management [5], as today defined by several standards like [17, 18], uses a quite large set of different views and perspectives for managing current IT. An effective architecture management approach for digital enterprises should additionally support the digitization of products and services [2] and be both holistic and easily adaptable [3, 13]. Furthermore, a digital architecture sets the base for the digital transformation enabling new digital business models and technologies that are based on a large number of micro-structured digitization systems with their own micro-granular architectures [3] like IoT [9, 10], mobile devices, or with Microservices [11, 12].

We are extending our previous service-oriented enterprise architecture reference model for the context of digital transformation with micro-granular structures considering associated multi-perspective architectural decision-making [19], which is supported by functions of an architectural cockpit [20]. Enterprise Services Architecture Reference Cube (ESARC) provides an architectural reference model [3] by bottom-up integrating dynamically composed micro-granular architectural models (Fig. 1). ESARC for digital products and services is more specific than existing architectural standards of architecture management, like in [17, 18].

ESARC [3, 13] uses eight integral architectural domains to provide a holistic classification model. Currently, it is still abstract from a concrete business scenario or technologies, because it is applicable for concrete architectural instantiations to support digital transformations [7, 8, 13]. The Open Group Architecture Framework TOGAF [19] provides the basic blueprint and structure for extended service-oriented enterprise architecture domains. Metamodel extensions are additionally provided by considering and integrating ArchiMate Layer models from [18].

Metamodels and their architectural data are the core part of the enterprise architecture. Enterprise architecture metamodels [5, 21] should enable decision making [21] as well as the strategic and IT/business alignment. Three quality perspectives are important for an adequate IT/business alignment and are differentiated as: (i) IT system qualities: performance, interoperability, availability, usability, accuracy, maintainability, and suitability; (ii) business qualities: flexibility, efficiency, effectiveness, integration and coordination, decision support, control and follow up, and organizational culture; and finally (iii) governance qualities: plan and organize, acquire and implement deliver and support, monitor and evaluate (e.g., [13]).

ESARC extends by a holistic view the metamodel-based extraction and bottom-up integration (Sect. 4) for micro-granular viewpoints, models, standards, frameworks and tools of a digital enterprise architecture model. ESARC frames these multiple elements of a digital architecture into integral configurations of an digital architecture by providing an ordered base of architectural artifacts for associated multi-perspective decision processes.

Architecture governance, as in [22], defines the base for well aligned management practices through specifying management activities: plan, define, enable, measure, and control. Digital governance should additionally set the frame for digital strategies, digital innovation management, and Design Thinking methodologies. The second aim of governance [23] is to set rules for a value-oriented architectural compliance based on internal and external standards, as well as regulations and laws. Architecture

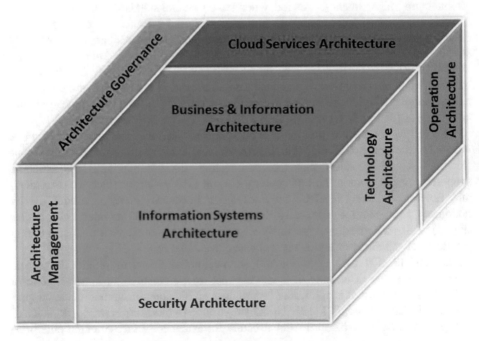

Fig. 1. Enterprise services architecture reference cube [3, 13]

governance for digital transformation [24] changes some of the fundamental laws of traditional governance models to be able to manage and openly integrate a plenty of diverse micro-granular structures, like Internet of Things or Microservices.

4   Architectural Composition

Digital transformation [1, 6, 7] not only changes our personal lives but also has massive implications on the competitive landscape. To win in this new environment, established companies need to develop new digitized products and services quickly, interact across channels, analyse customer behaviour in real-time, and leverage digital processes. Digitization can lower entry barriers for new players but causing long-understood boundaries between sectors to become more ambiguous and permeable. The nature of digital assets disaggregates value chains, creating openings for focused, fast-moving competitors.

Adaptability for architecting open micro-granular systems like Internet of Things or Microservices is mostly concerned with heterogeneity, distribution, and volatility. It is a huge challenge to continuously integrate numerous dynamically growing open architectural models and metamodels from different sources into a consistent digital architecture. To address this problem, we are currently formalizing small-decentralized mini-metamodels, models, and data of architectural microstructures, like Microservices and IoT into DEA-Mini-Models (Digital Enterprise Architecture Mini Model).

In general, such DEA-Mini-Models [11] consists of partial DEA-Data, partial DEA-Models, and partial EA-Metamodel. Microservices are associated with DEA-Mini-Models and/or objects from the Internet of Things [3]. Our model structures (Fig. 2) are extensions of the Meta Object Facility (MOF) standard [25] of the Object Management Group (OMG).

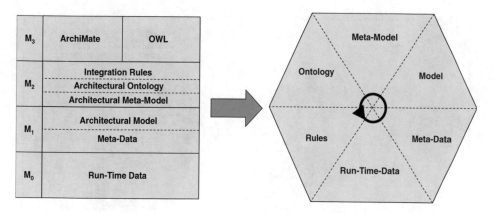

Fig. 2. Structure of EA-mini-descriptions [13]

Basically, we have extended the base model layer M1 to be able to host additionally metadata. Additionally, we have associated the original metamodel from layer M2 with our architectural ontology with integration rules. In this way we provide a close associated semantic-oriented representation of the metamodel to be able to support automatic inferences for detecting model similarities, like model matches and model mappings during runtime.

Regarding the structure of EA-Mini-Descriptions, the highest layer M3 [11] represents an abstract language concepts used in the lower M2 layer. It can be also seen as the meta-metamodel layer. The following layer M2 is the metamodel integration layer. The layer defines the language entities for M1 (e.g. models from UML or ArchiMate [18]). The models can be seen as a structured representation of the lowest layer M0 [25].

Volatile technologies, requirements, and markets typically drive the evolution of business and IT services. Adaptation is a key success factor for the survival of digital enterprise architectures [2, 3], platforms, and application environments. Weil and Woerner introduces in [6] the idea of digital *ecosystems* that can be linked with main strategic drivers for system development and system evolution. Reacting rapidly to new technology and market contexts improves the fitness of such adaptive ecosystems. Being a bit closer to the architecture and design of systems, Trojer et al. coined in [26] the *Living Models* paradigm that is concerned with the model based creation and management of dynamically evolving systems. Adaptive Object-Modelling and its patterns and usage provide useful techniques to react to changing user requirements, even during the runtime of a system. Moreover, we have to consider model conflict resolution approaches to support automated documentation of digital architectures and to summarize integration foundations for federated architectural model management.

During the integration of DEA-Mini-Models as micro-granular architectural cells (Fig. 3) for each relevant object, e.g., Internet of Things object or Microservice, the

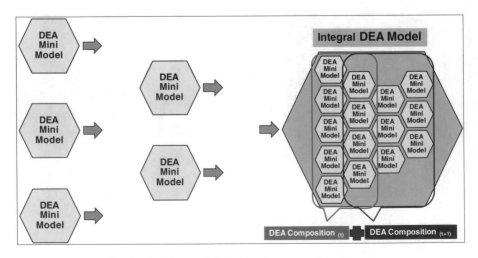

Fig. 3. Architectural federation by composition [3, 11]

step-wise composed time-stamp dependent architectural metamodel becomes adaptable [3, 11]. Furthermore, it can be mostly be automatically synthesized by respecting the integration context from a growing number of previous similar integrations [3].

In case of new integration patterns, we have to consider additional manual support. Currently, the challenge of our research is to federate these DEA-Mini-Models to an integral and dynamically growing DEA model and information base by promoting a mixed automatic as well as collaborative decision process, introduced and developed by Jugel in [19, 20], as in the following Sect. 5.

The Enterprise Services Architecture Model Integration (ESAMI) [13] (see Fig. 4) method is based on correlation analysis, which provides an instrument for a systematic manual integration process. Typically, this process of pair wise mappings is of quadratic complexity. We have linearized the complexity of these architectural mappings by introducing a neutral and dynamically extendable architectural reference model, which is supplied and dynamically extended from previous mapping iterations. Furthermore, we have adopted modeling concepts from ISO/IEC 42010 [27], like *Architecture Description*, *Viewpoint*, *View*, and *Model*.

The *Correlation Index* for different IoTs or microservices (red middle columns) with respect to the current *Reference Model* (yellow columns on the left) is created. Based on these *Correlation Indices*, the *Integration Options* for each service (green columns on the right) are chosen and the selection is integrated into the *Reference Model*. This continuous model refinement allows to integrate even extremely heterogeneous microservices that may not even share a complete metamodel.

These architectural metamodels are composed of their elements and relationships and are represented by architecture diagrams. The ESAMI approach is based on special correlation matrices, which are handled by a manual process to identify similarities between analyzed model elements. The chosen elements are then integrated according to their most valuable contribution towards a holistic reference model. In each iteration of this bottom-up approach, we are analyzing the fit of each new microservice metamodel in comparison with the context of the existing integrated set of services' metamodels.

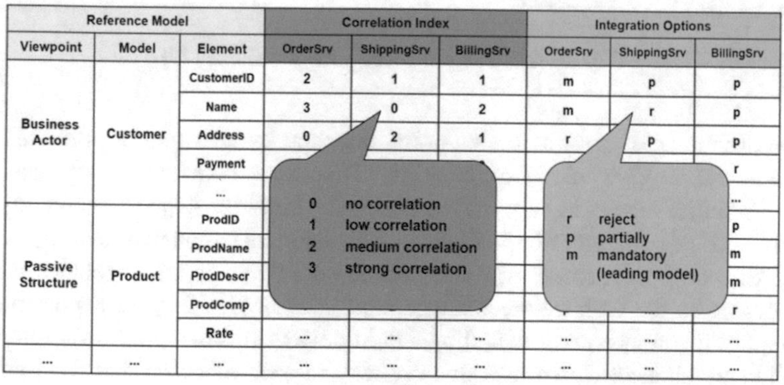

Reference Model			Correlation Index			Integration Options		
Viewpoint	Model	Element	OrderSrv	ShippingSrv	BillingSrv	OrderSrv	ShippingSrv	BillingSrv
Business Actor	Customer	CustomerID	2	1	1	m	p	p
		Name	3	0	2	m	r	p
		Address	0	2	1	r	p	p
		Payment						r
		...						...
Passive Structure	Product	ProdID						p
		ProdName						
		ProdDescr						m
		ProdComp						m
		Rate	...	...	...	...	...	...
...	...	...	...	...	...	...	...	...

0   no correlation
1   low correlation
2   medium correlation
3   strong correlation

r   reject
p   partially
m   mandatory (leading model)

Fig. 4. Correlation analysis and integration matrices [13]

5   Decision Management

Our current research links decision objects and processes to multi-perspective archi-tectural models and data. We are extending the more fundamentally approach of decision dashboards for Enterprise Architecture [5, 21, 23] and integrate this idea with an original Architecture Management Cockpit [19, 20] for the context of decision-oriented digital architecture management for a huge amount of micro-granular architectural models from the open-world.

As shown in Fig. 5, the architectural cockpit [19, 20] enables analytics as well as optimizations using different multi-perspective interrelated viewpoints on the system under consideration [3]. Multiple perspectives of architectural models and data result from a magnitude of architectural objects, which are typed according the dimension categories of a digital enterprise architecture from Sect. 2. Additionally, we have to consider analytics and decision viewpoints in a close association with the architectural core information.

The ISO Standard 42010 [27] defines, how the architecture of a system can be documented through architecture descriptions. Jugel et al. [20] develops and introduces a special annotation mechanism adding additional needed knowledge via an architec-tural model to an architecture description.

The advantage of architectural decision mechanisms is a close link between architectural artefacts and architectural models with explicit decisions, both from a classical Enterprise Architecture Management perspective and a new way of managing micro-granular structures and systems as well.

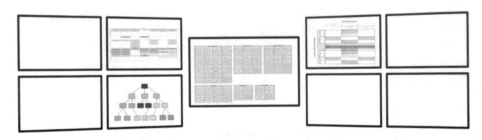

Fig. 5.   Architecture management cockpit [20]

In addition, [19] reveals a viewpoint concept by dividing it into an Atomic Viewpoint and a Viewpoint Composition. Therefore, coherent viewpoints can be applied simultaneously in an architecture cockpit to support stakeholders in decision-making [20]. Figure 6 illustrates the decision metamodel, as extension of [28], showing the conceptual model of main decisional objects and their relationships.

According to the architecture management cockpit [19, 20], each possible stake-holder can utilize a viewpoint that shows the relevant information. Furthermore, these viewpoints are connected in a dynamically way to each other, so that the impact of a change performed in one view can be visualized in other views as well.

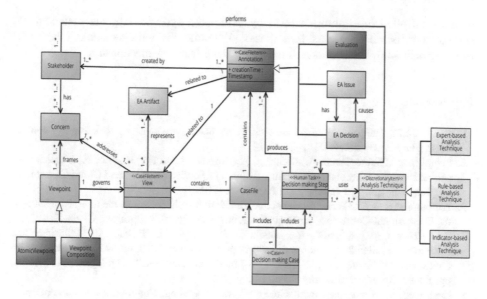

Fig. 6. Architecture decision metamodel [19]

6 Conclusion

Regarding our research question, we have first set the architectural background for digital services as well as digital products by focusing on main digitization concepts. Second, we have showed the need for an extended understanding and support of micro-granular systems as well as architectural models, like Internet of Things and Microservices.

The bottom-up composed digital enterprise architecture is a living digital enterprise architecture composition, which is in line with adaptive models and digital transformation mechanisms. This aspect of living architectural models fundamentally extends existing quite static standard frameworks like MODAF [29]. Strength of our research results from our novel integration of micro-granular structures and systems, while limits are still resulting from an ongoing validation of our research in practice and open issues of managing inconsistencies and semantic dependencies.

Our research question has pointed to a new viewpoint of an architectural composition, supported by a multi-perspective architecture management and decision environment for micro-granular digital architectures. Our novel main outcomes result from specific methods, mechanisms and environments for a decision-supported bottom-up integration for a huge amount of open-world micro-granular architectural structures as extended artifacts of a new tailored multi-perspective digital enterprise architecture. Furthermore, we are currently working on an extended architectural cockpit for digital enterprise architectures, related (engineering) processes using different extended decision support mechanisms.

Future research should be in the field of mechanisms for flexible and adaptable integration of digital enterprise architectures. Similarly, it may be of interest to extend human-controlled integration decision by automated systems, e.g. via mathematical comparisons (similarity, Euclidean distance), ontologies with semantic integration rules, or data analytics and data mining with deep learning mechanisms.

References

1. Porter, M.E., Heppelmann, J.E.: How smart connected products are transforming competition. Harvard Bus. Rev. **92**, 1–23 (2014)
2. Schmidt, R., Zimmermann, A., Möhring, M., Nurcan, S., Keller, B., Bär, F.: Digitization–perspectives for conceptualization. In: Advances in Service-Oriented and Cloud Computing, pp. 263–275. Springer International Publishing (2016)
3. Zimmermann, A., et al.: Multi-perspective digitization architecture for the Internet of Things. In: International Conference on Business Information Systems. Springer, Cham (2016)
4. Zimmermann, A., Schmidt, R., Sandkuhl, K., Jugel, D., Bogner, J., Möhring, M.: Decision-controlled digitization architecture for the Internet of Things and microservices. In: Czarnowski, I., Howlett, R., Jain, L.C. (eds.) Intelligent Decision Technologies 2017, Part II, pp. 82–92. Springer International Publishing (2018)
5. Lankhorst, M.: Enterprise Architecture at Work: Modelling, Communication and Analysis. Springer, Heidelberg (2017)
6. Weill, P., Woerner, S.: Thriving in an increasingly digital ecosystem. MIT Sloan Manag. Rev. **56**, 27–34 (2015)
7. Westerman, G., Bonnet, D.: Revamping your business through digital transformation. MIT Sloan Manag. Rev. (2015)
8. Brynjolfsson, E., McAfee, A.: The Second Machine Age: Work, Progress, and Prosperity in a Time of Brilliant Technologies. W. W. Norton & Company, New York (2014)
9. Atzori, L., Iera, A., Morabito, G.: The Internet of Things: a survey. Comput. Netw. **54**(15), 2787–2805 (2010)
10. Uckelmann, D., Harrison, M., Michahelles, F.: Architecting the Internet of Things. Springer, Heidelberg (2011)
11. Bogner, J., Zimmermann, A.: Towards integrating microservices with adaptable enterprise architecture. In: Dijkman, R., Pires, L.F., Rinderle-Ma, S. (eds.) IEEE – EDOC Conference Workshops EDOCW 2016 Vienna, pp. 158–163. IEEE (2016)
12. Newman, S.: Building Microservices: Designing Fine-Grained Systems. O'Reilly (2015)
13. Zimmermann, A., et al.: Architectural decision management for digital transformation of products and services. Complex Syst. Inform. Model. Q. **6**, 31–53 (2016)
14. Provost, F., Fawcett, T.: Data Science for Business: What You Need to Know about Data Mining and Data-analytic Thinking, 1st edn. O'Reilly Media, Sebastopol (2013)
15. Tiwana, A.: Platform Ecosystems: Aligning Architecture, Governance, and Strategy. Morgan Kaufmann, Amsterdam (2013)
16. Baldwin, C.Y., Woodard, C.J.: The architecture of platforms: a unified view. In: Platforms, Markets and Innovation, pp. 19–44 (2009)
17. Open Group: TOGAF Version 9.1. Van Haren Publishing (2011)
18. Open Group: ArchiMate 3.0 Specification. Van Haren Publishing (2016)
19. Jugel, D., Schweda, C.M., Zimmermann, A.: Modeling decisions for collaborative enterprise architecture engineering. In: 10th Workshop Trends in Enterprise Architecture Research (TEAR), held on CAISE 2015, Stockholm, Sweden, pp. 351–362. Springer (2015)

20. Jugel, D., Schweda, C.M.: Interactive functions of a Cockpit for enterprise architecture planning. In: International Enterprise Distributed Object Computing Conference Workshops and Demonstrations (EDOCW), Ulm, Germany, pp. 33–40. IEEE (2014)
21. Saat, J., Fanke, U., Lagerström, R., Ekstedt, M.: Enterprise architecture meta models for IT/Business alignment situations. In: IEEE-EDOC Conference 2010, Vitoria, Brazil (2010)
22. Weill, P., Ross, J.W.: It Governance: How Top Performers Manage It Decision Rights for Superior Results. Harvard Business School Press, Boston (2004)
23. Op't Land, M., Proper, H.A., Waage, M., Cloo, J., Steghuis, C.: Enterprise Architecture – Creating Value by Informed Governance. Springer, Heidelberg (2009)
24. Zimmermann, A., Jugel, D., Sandkuhl, K., Schmidt, R., Bogner, J., Kehrer, S.: Multi-perspective decision management for digitization architecture and governance. In: Dijkman, R., Pires, L.F., Rinderle-Ma, S. (eds.) IEEE – EDOC Conference Workshops EDOCW 2016 Vienna, pp. 174–181. IEEE (2016). ISBN 978-1-4673-9933-3
25. Object Management Group: OMG Meta Object Facility (MOF). Core Specification, Version 2.5 (2011)
26. Trojer, T., et al.: Living modeling of IT architectures: challenges and solutions. In: Software, Services, and Systems 2015, pp. 458–474 (2015)
27. Emery, D., Hilliard, R.: Every architecture description needs a framework: expressing architecture frameworks using ISO/IEC 42010. In: IEEE/IFIP WICSA/ECSA, pp. 31–39 (2009)
28. Plataniotis, G., De Kinderen, S., Proper, H.A.: EA anamnesis: an approach for decision making analysis in enterprise architecture. Int. J. Inf. Syst. Model. Des. 4(1), 75–95 (2014)
29. The British Ministry of Defence Architecture Framework Views. https://www.gov.uk/guidance/mod-architecture-framework. Accessed 24 Feb 2018

Revenue Management Information Systems for Small and Medium-Sized Hotels: Empirical Insights into the Current Situation in Germany

Michael Möhring[1]([⊠]), Barbara Keller[1], Rainer Schmidt[1],
and Alfred Zimmermann[2]

[1] Munich University of Applied Sciences, Lothstrasse 64,
80335 Munich, Germany
{michael.moehring, barbara.keller,
Rainer.Schmidt}@hm.edu
[2] Reutlingen University, Alteburgstraße 150,
72762 Reutlingen, Germany
alfred.zimmermann@reutlingen-university.de

Abstract. Revenue management information systems are very important in the hospitality sector. Revenue decisions can be better prepared based on different information from different information systems and decision strategies. There is a lack of research about the usage of such systems in small and medium-sized hotels and architectural configurations. Our paper empirically shows the current development of revenue information systems. Furthermore, we define future developments and requirements to improve such systems and the architectural base.

Keywords: Revenue management · Decision management · Architecture
Hospitality

1 Introduction

Revenue Management is important for Hotels to allocate the best room capacity to the right customer at the right price to ensure a high value of completeness and maximization of revenues [8]. In general, tourism enterprises like hotels offer services to a complex services system with different services operands and resources [9, 10]. The use of information technology is very important for hotels to produce competitive advantages [10, 11], to fulfil these services and implement a good revenue management [12]. Revenue management can be implemented by simple Excel-based spreadsheet tools or advanced systems [12]. Through digitization [13] more and more data sources are available to determine customer behaviour and set up the best prices to increase revenue. Besides large hotel chains like Hilton, small and medium-sized hotels have a lack in using advanced revenue management information systems, up to the current literature and our knowledge in practice. Therefore, there is a need to investigate the current situation of the use of revenue management information systems by small- and

© Springer International Publishing AG, part of Springer Nature 2019
I. Czarnowski et al. (Eds.): KES-IDT 2018, SIST 97, pp. 120–127, 2019.
https://doi.org/10.1007/978-3-319-92028-3_12

medium-sized hotels and show further possibilities for building-up advanced systems as well as change the information system's architecture.

This paper is part of an on-going research project answering the following research question: *Which is the current situation of using revenue information systems by small- and medium-sized hotels and how can this situation be improved?*

The remainder of the paper is structured as follows: After the introduction section, we define basic background information and the research method. In section four, we show the results of our single-case study and derive possibilities for revenues management system improvement. Finally, we conclude our research and show further research ideas.

2 Background

Revenue management tries to increase the profits of the supplier [18, 19]. Research reveals, that pricing and the associated concepts of pricing are important subjects in various research fields, like for example marketing and information systems (e.g., [20, 21, 23]). Due to the proceeding possibilities of digitization [13] and the coherent flexibility dynamic pricing approaches (e.g., [21, 25]), which are a partial field of revenue management [25], dynamic pricing gaining more and more attention in various business sectors. The hotel industry is no exception. Dynamic pricing can help to increase profitability [19], because it can help to implement a price discrimination [21]. Especially in the hotel industry such pricing concepts are quite interesting because of peculiarities like non-stock able services [12]. Furthermore, rooms have fixed costs not depending on the occupancy of the room [22]. That means, for a hotel it is much better to sell a room for a lower price than not to rent out the room, because there are still costs [12].

Furthermore, it is not possible to store the service (here: e.g. room capacity) and compensate the costs later [12]. In the hotel industry, selling identical rooms to comparable customers for different fees can be seen as a typical example for the implementation of dynamic pricing [19]. A lot of clarifying examples can be mentioned in this context. For instance, booking a hotel for the summer holidays a long time in advance might be cheaper. Because if the occupancy of the hotel rooms increases due to a high demand, the last rooms will be sold to persons with a high price acceptance. Decision support systems try to support suppliers and address this challenge with forecasts based on historical data [24]. Therefore, hotel revenue management might be defined with regards to Ivanova et al. [19] as a composition of different tools and measures focussing on the optimization of net gained revenues by giving the right offer to the right customer at the right time on the right channel. In this case, demand and supply respectively capacity and occupancy determine an actual dynamical adjusted price level [19].

A lot of hotels are still using a form of dynamic pricing to increase their revenues [25]. However, the way they implement it is quite heterogeneous. Nevertheless, at last they have one important factor in common, which they want to address with their dynamic pricing efforts: the costs. The implementation of a dynamic pricing system helps to gain more profit without leading simultaneously to a high amount of costs [26].

The information systems architecture of a hospitality enterprise consists of different, heterogenous information systems [1] (e.g., booking platform systems, ERP-systems, systems of online-travel agencies (OTAs), revenue information systems). According to current research, there is a lack of architectural research in this area [1]. Furthermore, more and more data-intensive processes have to support the enterprise architecture [2]. Nevertheless, there is a sparse research about the architecture of revenue information systems for small and medium-sized hotels and hotel chains according to a literature review [3] in leading databases like AiSel, IEEExplore, Sciencedirect, Springer link with keywords related search items.

Revenue management information systems are quite important for small-and medium-sized enterprises. But there is a lack in research about the current situation and further developments as described above. For answering our research question, we used a single case study method described in the following.

3   Research Method and Data Collection

We used a single-case study research method to investigate our research question according to [4]. Single case studies can generate interesting insights and generable empirical circumstances [5]. Furthermore, information systems research is often using single case studies for investigations (e.g., [6, 7, 17]).

The observed hotel is a four-star superior luxury hotel in Germany with different restaurants, wellness and SPA opportunities. Furthermore, the hotel will build up new hotels and the management is well connected to the German hospitality association (focussing on small and medium-sized hotel (chains)). The qualitative data (interview data) were collected in quarter four of the year 2017 and analysed regarding coding technique recommended by the literature [4, 16].

To ensure a high quality of research and the validity of our results, we further discuss the aspects with leading OTAs and booking platforms in Europe. In the following section, we show the results.

4   Results and Further Development

In the observed luxury hotel and in line with discussions with OTAs and booking platforms, revenue management is implemented with an easy to use revenue management software tool without any analytics support or screening functions. The management told us, that this is the same case for the most of small and medium-sized hotels according to insights of the hospitality association.

One year ago, there was no revenue management tool used by the management of this hotel. The tool is connected to the different OTAs as well as the website of the hotel to show all customers the defined price for the type of room for the specific time. In the past, every day (without some special days like Christmas or weekend specials) there was always the same room rate. After using this revenue management tool, revenues were increased. Different prices lead to a better occupancy rate, because different and further persons could be attracted. For instance, people with less time restrictions could find out when prices are lower and used that opportunities.

The hotel management marked out in this context, that at first they were afraid to implement the dynamic price approach. They were not sure how customers would react to the flexible adjustment of the prices. Especially, the risk that customers could maybe remember a former lower price seemed to be high for them as a medium-sized hotel. Nevertheless, they decided for themselves to take the risk. Surprisingly, the customers did not react in a negative manner. If questions arrived, they focussed in their reasoning on the positive aspects for the customers (e.g., could gain a better price). Furthermore, they told us that they implemented an additional step for their steady customers (5 to 10 families) returning several times a year. They give them the choice between two options. In the first option, the steady customers could also be participating in the dynamic pricing. That means, that they get the actual and adjusted price like all the others. In the second option, they could have the usual price from the past also in the future, but the price stays static. That means, there will be no adjustment of the price in anyway. In consequence, the customers have to pay always the same price, independent of the current hotel room vague. Doing so, no customers did negatively react to the implementation of dynamic prices.

Besides the customer-related aspects mentioned above, the information system architecture enabling the implementation of such revenue management field must be considered. Also, this more technological aspect is very challenging for small and medium-sized hotels and hotels chains. Typically, we have to consider an integral architectural viewpoint [27], which integrates different aspects of the enterprise architecture: from the strategy and a decisional IT-governance support over the business architecture, to the information services, and the technology architecture, all on base of a clear modelled security and data architecture, with methods and instruments for data analytics. The architecture of our case is shown in the following Figure (Fig. 1):

Fig. 1. Architecture

Regarding, the optimization driver for revenue management in hospitality [12] such simple tools and the current revenue management system architecture can not support a

Table 1. Used and implemented optimization driver of the case.

Optimization driver according to [12]	Short description	Fulfilled by the implemented revenue management information system and used in our case
Quota restriction	Define quotas to a defined number of sales units/price levels [12]	Medium (only based on the room type or special events/discounts)
Overbooking	Accepting more booking based on the analysis of no-shows [12]	No
Itinerary or length of stay management	Considering the different booking length of customers [12]	No
Optimization of the distribution network	Reducing distribution costs, choosing a good distribution network [12]	Medium (only static price exchange to the connected OTAs)

good and competitive revenue management (Table 1). Most of the optimization drivers are not support by the systems in our case.

Therefore, revenue management information systems for small and medium-sized hotels should better apply and integrate such optimization drivers to increase revenues. For the implementation a better forecasting of average prices, occupancy rates [12] and scanning the average prices of the competitor hotels are needed to implement. Small and medium-sized hotels have often not a high capability in the information systems field and related experts for IT (like in our case study). Therefore, revenue information systems should be easy to use and implemented in a user-friendly scalable environment.

Table 2. Further data sources needed based on our case

System	Description	Data format
OTA	Collecting price and available rooms for own and competing hotels	Structured
Hotel website	Collecting booking and search behaviour	Mostly structured
Partner website	Collecting price and available rooms for competing hotels	Structured
Competitor website	Collecting price and available rooms for competing hotels	Mostly unstructured (Collection via Web-Crawling)
Social media	Collecting data about booking behaviour and interest	Mostly unstructured data (e.g., opinions and travel discussions for hotels, regions, trends, etc.)
Google trends	Collecting data about booking behaviour depending on search behaviour	Structured data (Source for price and occupancy forecast based on Google search volume)

Furthermore, forecasting and calculating the available rooms and prices, quotas, etc. can be very complex [12]. Therefore, revenue management information systems implemented in a public cloud [2, 14] environment could be one solution to fulfil this gap and save unused capacity and related costs. The acceptance of business users of such a public cloud solution is increasing and depending on different factors [15].

Furthermore, different information systems should be integrated (in general as well as in our case) to ensure a good quality of price and occupancy forecasting (Table 2).

Regarding Table 2, there is e.g. a need to collect pricing data as well as the number of available rooms from the different OTAs. Furthermore, also unstructured data (e.g., from social media) must be integrated. The hotel of our case as well as all small-and medium sized hotels should integrate and use this data sources to ensure a good quality of price and occupancy forecasting.

In the following a conclusion of our research and future directions are shown.

5 Conclusion and Future Directions

Our study has found a fundamental research gap about revenue management systems for small-and medium-size hotels. Furthermore, also practice shows a huge gap of implementing such tools for improving revenues, decisions and competitiveness. Although we found out during our case study that even small and medium-sized hotels and hotel chains can profit from dynamic pricing systems. Based on a case study, we have analysed the current situation and set further improvements for a future information systems architecture and a revenue management decision support system.

Our research contributes in different ways. First, research can benefit from new knowledge about revenue management information systems in small- and medium-sized hotels. Architectural components and future data-driven possibilities and implications were defined. Second, (revenue) managers of small- and medium-sized hotel chains can use our research to evaluate and further develop their revenue management approaches and information systems architecture.

Our research builds on a single case. But the insights are in line with discussions with OTAs and the interviewed managers are well connected to the hospitality association and its members (e.g., other small- and medium-sized hotels). Future research should therefore add more cases from different countries. Furthermore, more qualitative and quantitative studies are needed to get more insights in special aspects of revenues information systems for small and medium-sized hotels (e.g., OTA data exchange improvement to protect undesired overbooking).

References

1. Schmidt, R., Möhring, M., Keller, B., Zimmermann, A., Toni, M., Di Pietro, L.: Digital enterprise architecture management in tourism–state of the art and future directions. In: International Conference on Intelligent Decision Technologies, pp. 93–102 (2017)

2. Keller, B., Möhring, M., Toni, M., Di Pietro, L., Schmidt, R.: Data-centered platforms in tourism: advantages and challenges for digital enterprise architecture. In: International Conference on Business Information Systems, pp. 299–310 (2016)
3. Webster, J., Watson, R.T.: Analyzing the past to prepare for the future: writing a literature review. MIS Q. **26**(2), pp. xiii–xxiii (2002)
4. Yin, R.: Case Study Research: Design and Methods. Sage Publishing, Beverly Hills (1994)
5. Darke, P., Shanks, G., Broadbent, M.: Successfully completing case study research: combining rigour, relevance and pragmatism. Inf. Syst. J. **8**, 273–289 (1998)
6. Warrington, E.K.: The fractionation of arithmetical skills: a single case study. Q. J. Exp. Psychol. **34**, 31–51 (1982)
7. Walsham, G., Waema, T.: Information systems strategy and implementation: a case study of a building society. ACM Trans. Inf. Syst. (TOIS) **12**, 150–173 (1994)
8. Kimes, S.E.: The basics of yield management (1989)
9. Vargo, S., Lusch, R.: Service-dominant logic: continuing the evolution. J. Acad. Mark. Sci. **36**, 1–10 (2008)
10. Berne, C., Garcia-Gonzalez, M., Mugica, J.: How ICT shifts the power balance of tourism distribution channels. Tour. Manage. **33**, 205–214 (2012)
11. Buhalis, D., Amaranggana, A.: Smart tourism destinations enhancing tourism experience through personalisation of services. In: Information and Communication Technologies in Tourism, pp. 377–389 (2015)
12. Legohérel, P., Fyall, A., Poutier, E.: Revenue management for hospitality and tourism (2013)
13. Schmidt, R., Zimmermann, A., Nurcan, S., Möhring, M., Bär, F., Keller, B.: Digitization – perspectives for conceptualization. In: Celesti, A., Leitner, P. (eds.) Advances in Service-Oriented and Cloud Computing, ESOCC Workshops 2015, Taormina, Italy. Springer, Cham (2016)
14. Mell, P., Grance, T.: The NIST definition of cloud computing (2011)
15. Schmidt, R., Möhring, M., Keller, B.: Customer relationship management in a public cloud environment–key influencing factors for european enterprises. In: Proceedings of the 50th Hawaii International Conference on System Sciences (2017)
16. Strauss, A., Corbin, J.: Open Coding. Basics of Qualitative Research: Grounded Theory Procedures and Techniques, vol. 2, pp. 101–121. SAGE Publications, Thousand Oaks (1990)
17. Laumer, S., Maier, C., Eckhardt, A.: The impact of human resources information systems and business process management implementations on recruiting process performance: a case study (2014)
18. Heo, C.Y., Lee, S.: Influences of consumer characteristics on fairness perceptions of revenue management pricing in the hotel industry. Int. J. Hospitality Manag. **30**(2), 243–251 (2011)
19. Ivanova, M., Ivanov, S., Magnini, V.P.: The Routledge Handbook of Hotel Chain Management. Routledge, New York (2016)
20. Dodds, W.B., Monroe, K.B., Grewal, D.: Effects of price, brand, and store information on buyers' product evaluations. J. Mark. Res. **28**(3), 307–319 (1991)
21. Hinz, O., Hann, I.-H., Spann, M.: Price discrimination in e-commerce? An examination of dynamic pricing in name-your-own price markets. MIS Q. **35**(1), 81–98 (2011)
22. Zeithaml, V.A.: Consumer perceptions of price, quality, and value: a means-end model and synthesis of evidence J. Mark. **52**(3), 2–22 (1988)
23. Guo, X., Ling, L., Yang, C., Li, Z., Liang, L.: Optimal pricing strategy based on market segmentation for service products using online reservation systems: an application to hotel rooms. Int. J. Hospitality Manag. **35**, 274–281 (2013)

24. Guadix, J., Cortés, P., Onieva, L., Muñuzuri, J.: Technology revenue management system for customer groups in hotels. J. Bus. Res. **63**(5), 519–527 (2010)
25. Abrate, G., Fraquelli, G., Viglia, G.: Dynamic pricing strategies: evidence from European hotels. Int. J. Hospitality Manag. **31**(1), 160–168 (2012)
26. Bapna, R., Goes, P., Gupta, A., Jin, Y.: User heterogeneity and its impact on electronic auction market design: an empirical exploration. MIS Q. **28**(1), 21–43 (2004)
27. Zimmermann, A., Schmidt, R., Sandkuhl, K., Jugel, D., Bogner, J., Möhring, M.: Decision-controlled digitization architecture for internet of things and microservices. In: International Conference on Intelligent Decision Technologies, pp. 82–92 (2017)

On Knowledge-Based Design Science Information System (DSIS) for Managing the Unconventional Digital Petroleum Ecosystems

Shastri L. Nimmagadda[1], Neel Mani[2(✉)], and Torsten Reiners[1(✉)]

[1] School of Management, Information Systems, CBS, Curtin University,
Perth, WA, Australia
shastri.nimmagadda@curtin.edu.au,
T.Reiners@cbs.curtin.edu.au
[2] School of Computing, University College Dublin, Dublin, Ireland
neelmanidas@gmail.com

Abstract. The unconventional digital petroleum ecosystems are associated with fractured reservoirs that are usually unpredictable, but can produce for longer periods depending on size of petroleum systems and basins. Currently, conventional reservoirs do produce oil & gas even without integrated workflows and solutions. The heterogeneity and multidimensionality of data sources at times can make the data documentation and integration complicated affecting the exploration and field development. We examine the conventional database technologies and their failures in organizing the data of unconventional digital ecosystems. Big Data driven intelligent information system solutions are needed for addressing the issues of complex data systems of unconventional digital ecosystems. Geographically distributed petroleum systems and their associated reservoirs too demand such integrated and innovative digital ecosystem solutions. We propose an innovative design science information system (DSIS), an integrated digital framework solution to explorers, dealing with unconventional fractured reservoirs. The integrated Big Data analytics solutions are effective in interpreting unconventional digital petroleum ecosystems that are impacted by shale prospect businesses worldwide.

Keywords: Digital ecosystem · Domain ontologies · Big-data
Data integration · Data warehousing and mining · Data fusion
Unconventional data sources · Evaluation properties

1 Introduction

Worldwide sedimentary basins [5] comprise of numerous conventional and unconventional shale deposits. Each of these basins has the capability of generating and producing numerous unconventional oil and gas fields [6, 8] from multiple petroleum systems and basins. Each field has multiple oil and gas producing wells and each drilled-well has multiple unconventional reservoir pay zones, with each pay zone having different fluids - either oil and or gas. In a basin scale, this is a typical data hierarchy with

© Springer International Publishing AG, part of Springer Nature 2019
I. Czarnowski et al. (Eds.): KES-IDT 2018, SIST 97, pp. 128–138, 2019.
https://doi.org/10.1007/978-3-319-92028-3_13

existence of super-type and subtype dimensions that attribute to an oil*play* [1], a scalable and atomic property of a *prospect*. The source rocks are unconventional reservoirs. In a digital oil field scenario, the data sources interpreted as entities and or dimensions in multiple domains are logically organized through ontological descriptions to integrate in a data warehouse environment [12, 13]. The data in multiple unconventional oil and gas fields are organized in a way their connectivity can be established from multiple databases, using semantic tools and technologies. In the present research, issues associated with existing methodologies, significance and motivation of innovative approaches and how they address the challenging issues are discussed.

2 Previous Studies

There has been an explosive growth in the volumes of published petroleum data, as a result of the rapid exploration and development activities worldwide [5, 13]. Organization of such data by industry's professional organizations, national, private oil and gas companies including major service companies is a huge challenge. This has motivated us to organize and manage a wealth of data [2, 3, 13] accumulated in multiple sources linked to different ecosystems [13]. Several elements and processes involved in building a digital ecosystem are described in [11]. The feasibility and applicability [13] of the proposed technologies are analyzed for exploring the connectivity and the boundaries of system elements and processes. Use of ontologies, implementing them in a data warehouse approach is discussed comprehensively in [13, 14]. Use of sets and set theory in different applications are given in [7, 14]. How the principles of set theory can adapt to Big Data features are described in [12].

3 Issues and Challenges

Around the world, the unconventional gas resources are widespread, but with several exceptions that they have not received careful attention by natural gas operators, especially with lack of logistics for exploiting unconventional petroleum systems. In addition, there is a chronic shortage of expertise in specific technologies needed to develop the unconventional resources successfully. As a result, only limited development of technology has taken place outside of North America. An interest is however growing and during the last decade, efforts are being made to develop unconventional gas reservoirs worldwide. Integrating or assembling information from several databases to solve problems and discovering new knowledge are other major challenges in geo-informatics. Exchanging volumes and variety of exploration data into valuable geological knowledge is the challenge of knowledge discovery process. Data mining and interpretation of interesting patterns hidden in petabyte size of seismic and other exploration data are critical goals of geo-informatics. The goals cover identification of useful reservoir bearing structures from multiple petroleum ecosystems [11, 13] and their connectivity that are of commercial interest to explorers and investors.

In addition, other major issues are poorly managed upstream integration with multifaceted Big Data dimensions [4]. Besides, heterogeneity and multidimensionality of conventional data structures (relational, hierarchical and networking data types, [12, 13]) including complex spatial-temporal data [9] sources pose serious challenges. Similar issues [14] exist in exploration and production of unconventional Big Data sources. Massive storage devices are needed for volumes of business/technical data dimensions (instances to the petabyte size for each unconventional oil and gas field, an entity in a major producing basin) that describe oil and gas fields: for each field – number of *surveys* and number of *drilled-wells*; in each well, number of *horizons* (geological formations) and in each survey: number of survey lines, with hierarchical and relational ontology descriptions [14]. Handling numerous data dimensions (entities and or objects) to the order of thousands of attributes, mapping and modeling their dimension and fact tables are tedious process [10]. Data integration (sometimes among 50 to 80 geographically distributed operational centers with an Indonesian case study [17]) of multi-disciplinary data dimensions is a serious business issue in large oil and gas companies.

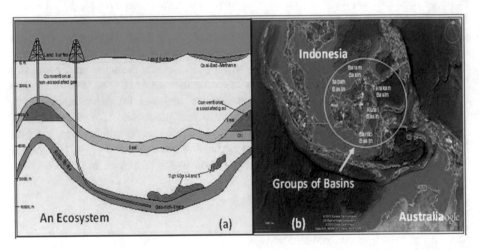

Fig. 1. Representation of (a) unconventional resources within a single petroleum ecosystem; (b) Google imagery view showing Southeast Asian Sedimentary basins and their inherent connectivity

As illustrated in Fig. 1, multiple sedimentary basins, associated petroleum systems and sets of oil and gas fields existing in a marine environment all linked together, constitute an ecosystem. In conventional petroleum ecosystem environment, the oil and gas can produce without much effort on exploration and field development because of types of reservoirs, and their qualities. In unconventional digital ecosystems, the reservoirs are unconventional and their petroleum sources are typically shales with network of fractures [2] and cannot produce on their own unless the shale rocks (petroleum-source) are stimulated. In the ecosystems view point, we are of the opinion that both conventional and unconventional systems share information and knowledge, all encapsulated in large size digital ecosystem [13]. This phenomenon opens new opportunities to develop smarter and knowledge base technologies in upstream petroleum industries.

4 Motivation and Significance

The resources needed in running large size integrated upstream projects are abundant. The heterogeneous and multidimensional Big Data sources that surround the upstream oil and gas companies motivate us to develop new data integration methodologies. Knowledge-based and smarter multidimensional data constructs and models are proposed for designing and developing an integrated design science information system (DSIS) framework [9, 16]. Besides, the conceptualization and contextualization attributes of various entities and dimensions that support the innovative methodology have further led to develop a multidimensional digital ecosystem-based inventory or repository. Additionally, the fine grained multidimensional data structuring with domain ontology descriptions makes the DSIS approach more smart and flexible in an environment, where varieties of business rules apply to models and when the constraints rapidly change [14] either in geology or upstream business. The visualization and interpretation are other significant artefacts of the Big Data facilitating the use, reuse, test the interoperability and effectiveness of the data models for sustainable DSIS including an effective data analytics [12, 14].

5 Methodology

Designing and developing an integrated digital framework have been the focus of current research in integrating intelligent digital ecosystem scenarios [7, 13, 14]. In worldwide unconventional shale-gas basins multiple petroleum systems (information systems) exist with variety of elements and processes and volumes of their fact instances. We examine these unstructured data sources for modelling unconventional digital petroleum ecosystems. The conventional petroleum resources inherently embed with unconventional shale gas, shale oil and coal bed methane (CBM) resources, all existing within a single digital ecosystem scenario. A robust methodology [3, 12] is vital for managing the digital ecosystem scenarios under the control of Big Data sources and their sizes. Design Science Information System (DSIS) strategy is an alternative approach, addressing the data modelling and integration challenges with heterogeneous and unstructured data sources. Big Data-driven DSIS is simulation of an ecosystem framework [9, 12] in which several research activities are articulated including implementation of research outcomes that are validated by various evaluation properties.

Unconventional digital ecosystems hold large volumes and varieties of data in multiple domains. For designing, developing and implementing the digital ecosystems in Big Data scale, high-quality and reliable data sources are necessitated. For the purpose of bringing multidimensional heterogeneous data into the integration process [13], several data instances are gathered from unconventional ecosystems. Hundreds of dimension- and -fact tables including attributes and their corresponding data instances are considered in the modeling purposes. These attributes are either hierarchically or relationally structured or both used for modeling the data instances in a metadata [14]. Several *point, line, spatial* (areal) dimensions are ontologically described and integrated in such a way data structures can establish connectivity between seismic- and well-domain datasets. Geographically metadata are mappable for fractured network models.

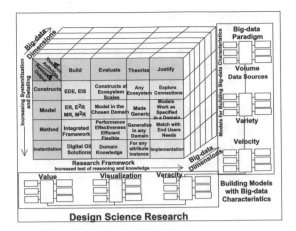

Fig. 2. Big data supportive DSIS Framework

For building data structures, the domain ontologies are described, supporting the DSIS warehouse repository. It can connect hundreds of logical multidimensional data relationships from digital ecosystems (Fig. 2). We reiterate that the concept of a digital ecosystem in the DSIS context is to bring multiple data dimensions together and connect with manifold multidimensional schemas. We incorporate the characteristics of Big Data logically in DSIS constructs and models, integrating several multidimensional warehouse repositories into a meta-metadata. Domain ontologies search for connections between disconnected reservoirs through interconnected multidimensional data relationships and their associated schemas. Various algorithms are described in [14] demonstrating the artifacts including their integration process. Other artifacts in the form of ontology models that go into the DSIS are described in the following sections.

5.1   Time-Domain Ontologies

The multidimensional data models are in both time- and depth-domains, with ontological descriptions for every time- and -depth ranges. Information on lithology existing in different intervals of drilled wells and their time occurrences in the seismic data are incorporated in the multidimensional modelling process. The seismic exploration data are represented in different space and time dimensions and separate ontologies are written for multiple seismic vertical time ranges: 0–500 ms, 500–1000 ms, 1000–1500 ms, 1500–2000 ms and 2000–2500 ms. Each vertical time range has different geologically interpretable knowledge, such as horizons, structures, reservoirs, source and seal rocks, and other various petroleum system's processes. Data instances from a time-domain cross section are documented as shown in Fig. 3. This cross section covers a complex reservoir, interpreted from integrated seismic and drilled-well metadata.

On Knowledge-Based Design Science Information System (DSIS)   133

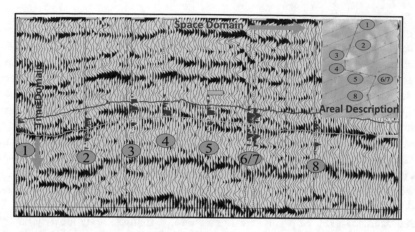

Fig. 3. Seismic *peak* and *trough* dimensions and their connectivity (time-domain ontologies)

5.2 Depth-Domain Ontologies

Similar knowledge based ontologies are described in depth-domain, in which different ranges 0–500 m, 500–1000 m, 1000–1500 m and 1500–2000 m are labeled (Fig. 4). Each range has different knowledge levels; for example, CBM occurs in the range of 0–500 m, conventional non-associated gas from 500–1000 m, conventional associated gas from 1000–1500 m and tight gas sands beyond 2000 m. Hydrocarbons associated within shale reservoir can be deeper (more than 3000 m), where in, they experience matured temperatures for hydrocarbon generation. Extraction and interpretation of knowledge for deeper reserves of shale gas requires skill and effort and it is more expensive to exploit.

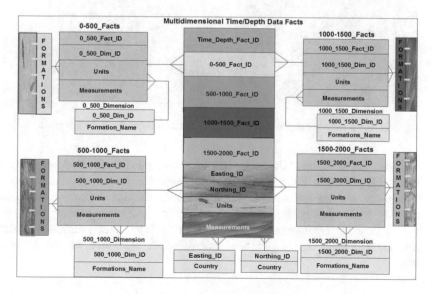

Fig. 4. Depth domain ontologies

5.3   Horizon-Based Ontologies

The seismic data possess multiple horizons with *peak* and *trough* dimensions in modeling either seismic times or computed depth structures. We believe a horizon (either seismic or geological) should never be viewed as a single entity or dimension nor isolated from multiple horizons at any stages of processing and or interpretation. Multiple horizons are ontologically interlinked or interconnected in an ecosystem scenario with conceptualized "velocity" attribute. Horizons in a sedimentary basin scale form a collective ecosystem, in which all the horizons interact and communicate (during several geological periods) among themselves through shared properties such as velocity, density including porosity or permeability of reservoirs. If there is any change in a single horizon, there is a corresponding change in other horizons (reasons could be structural connection and reservoir extensions in multiple horizons in 2D/3D seismic datasets) and their properties. In an analogy, multiple horizons may have been associated with multiple geological ages. A simulated fractured cross-section is shown in Fig. 5a. Each horizon is characterized by attribute dimensions of *structure, reservoir, seal* and *source* elements as illustrated in a dimensional model in Fig. 5b.

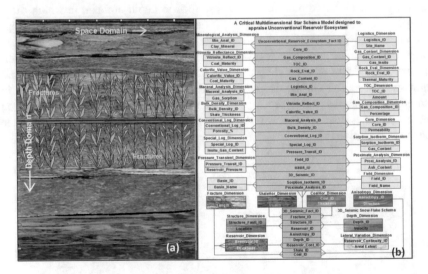

Fig. 5. (a) Fractured geological section (b) multidimensional data schema

One of the key multidimensional models representing the *shale-gas* exploration and development is presented in Fig. 5. Several interconnected dimensions which represent one-to-one, one-to-many and many-to-many data relationships in *exploration* domain, are logically (ontologically described) connected to their corresponding fact tables within a warehouse environment. For connecting elements of multiple petroleum systems, corresponding ontologies are articulated as described in the following sections.

Source Ontologies. The *source* rock is the main element in the ontology description. A rock rich in organic matter, if heated sufficiently can generate oil or gas. Typical source rocks, usually shales or limestones, contain about 1% organic matter and at least 0.5% total organic carbon (TOC), although a rich source rock might have as much as 10% organic matter. Preservation of organic matter without degradation is critical to creating a good source rock, and necessary for a complete petroleum system. Under right conditions, the source rocks can also be reservoir rocks, as in the case of shale gas exploration & production (E & P). Among several dimensions involved within source ontology, ontology description of maturity of source rock is a characteristic property of any petroleum ecosystem that can ascertain the potentiality of unconventional reservoirs.

Structure Ontologies. The *structure* and or entrapment of hydrocarbons [11] are other key elements of petroleum ecosystem. All the necessary inputs needed to explore the trapping mechanism are articulated from ontology descriptions through structural (geological) connectivity and corresponding data connectivity in the warehouse repository.

Seal Ontologies. The *seal* rock is another characteristic property of an entire petroleum ecosystem habitat. Unless an appropriate seal rock is interpreted on a regional scale, oil and gas field trapping mechanism cannot be well understood in its totality within an ecosystem setting. Accordingly the ontologies associated with seals described in [11] are integral part of the ecosystem to make connectivity between systems through seals.

Reservoir Ontologies. A reservoir is one of the principal components of a petroleum ecosystem. The reservoir potentiality demonstrates large accumulations of hydrocarbons in a basin. Our primary focus is modeling the reservoir connections using an integrated ontology framework that supports the multidimensional warehouse modeling and mining [14] in Big Data scale.

6   Evaluation and Implementation of DSIS Artefacts

Like elements of an ecosystem, the existing processes also affect the potentiality of the E & P. The elements and processes are together explored for their connections to a new conceptualized element, termed as {chains}. "Chains" within an ecosystem are more conceptualized and they are based on cognitive interaction between elements and processes. A Chain can either be between individual elements and or processes of petroleum systems from groups of basins. The seismic data that make up the chains consist of numerous *peak* and *trough* dimensions [14, 15]; they are used for connecting horizons and their associated elements structure, reservoir, source and seal of a petroleum ecosystem. The data associated with elements and processes are digitally modeled, as sets using set theory; cardinalities of elements are thus described in several permutations and combinations [14]. Then each element is narrated with an ontological description to formulate the connectivity. Description of ontologies for each individual element is, though a primary focus, establishing the reservoir connections and integrating the peak and trough dimensions (of seismic data) within warehoused metadata are other foci as shown in Fig. 6. The map view drawn from the meta-metadata corroborating the reservoir connections and exploring drillable prospects demonstrates the model validity in the upstream business contexts.

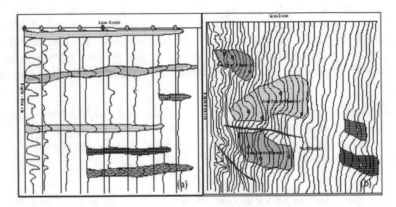

Fig. 6. Exploring reservoir connections in space and depth dimensions on well-log data (a) one dimensional map view and (b) 2D areal map view

7   Knowledge Management and Research Deliverables

The digital framework that characterizes the data connectivity between multiple domains and systems is a new innovative solution, integrating digital schemas in a unified multidimensional repository. The constructs, models, and methods are final deliverables of the current research application. Models and metadata cubes [7, 14] deduced from integrated framework (Fig. 2) facilitate us an effective and cognitive visualization and interpretation of digital petroleum ecosystems. The digital petroleum ecosystem framework ensures an interpretable metadata with predictive knowledge of unconventional petroleum systems as demonstrated in Fig. 6. Artefacts used in the DSIS are evaluated using various utility properties such as "data quality, data governance, interoperability, usefulness, ease of use and effectiveness", establishing the efficacy of DSIS and managing the unconventional digital petroleum ecosystems [14]. Overall, the DSIS has a vital role in making huge impacts in smarter integrated interpretation upstream projects, especially during *prospect identification* and *risk evaluation* stages.

8   Conclusions and Future Work

Big data characteristics may have inherited from volumes of unconventional petroleum data sources, meant for exploring connections between *shale gas*, *shale oil*, *coal bed methane* (CBM) and *tight oil* sands, all existing within a single sedimentary basin, an unconventional ecosystem. Unconventional data sources are heterogeneous and multidimensional. For data integration, ontology based data warehousing and mining is used. Besides, visualization and interpretation artifacts are part of knowledge-based smart digital ecosystem solutions. Ontologies written for multiple dimensions facilitate the connectivity among unconventional petroleum ecosystems. The Big Data technologies are invaluable tools for investigating, examining and interpreting the sweet-spots, besides smart delivery of fine-grained metadata cubes of several unconventional reservoir ecosystems that validate the connections among multiple reservoir systems.

References

1. Beaumont, E.A., Foster, N.H.: Exploring for Oil & Gas Traps. In: AAPG Treatise of Petroleum Geology, 2nd edn. Publications of Millennium Edition, Memoir 78, UK (1999)
2. Brown, D.: Looking Deeper into Fracture Impacts, AAPG Explorer, March Archive, USA (2013)
3. Carvajal, G., Maucec, M., Cullick, S.: Intelligent Digital Oil and Gas Fields, 1st edn. Gulf Professional Publishing, Elsevier, 374 p. (2017)
4. Cleary, L., Freed, B., Elke, P.: Big Data Analytics Guide. SAP, CA 94607, USA (2012)
5. Li, G.: World Atlas of Oil and Gas Basins. Wiley, New York, 496 p. (2011)
6. Castaneda, G.O.J., Nimmagadda, S.L., Cardona, M., Lobo, A., Darke, K.: On integrated quantitative interpretative workflows for interpreting structural and combinational traps for risk minimizing the exploratory and field development. In: Bolivarian Geophysical Symposium Proceedings, held in Cartagena, Colombia (2012)
7. Coronel, C., Morris, S.: Database Systems: Design, Implementation, & Management. Cengage Learning US, Edition 12, 784 p. (2016)
8. Durham, S.L.: An Unconventional Idea, Open to Interpretation. AAPG Explorer, March Series, USA (2013)
9. Indulska, M., Recker, J.C.: Design SCIENCE in IS research: a literature analysis. In: Gregor, S., Ho, S. (eds.) Proceedings 4th Biennial ANU Workshop on Information Systems Foundations, Canberra, Australia (2008)
10. Khatri, V., Ram, S.: Augmenting a conceptual model with geo-spatiotemporal annotations. IEEE Trans. Knowl. Data Eng. **16**(11), 1324–1338 (2004)
11. Magoom, L.B., Dow, W.G.: The Petroleum System from Source to Trap, AAPG/Datapages, Digital Reprint of AAPG Memoir 60 (2009)
12. Nimmagadda, S.L., Rudra, A.: Big data information systems for managing embedded digital ecosystems (EDE), a book chapter in a book entitled. In: Big Data and Learning Analytics in Higher Education: Current Theory and Practice. Springer, The Netherlands (2016). https://doi.org/10.1007/978-3-319-06520-5, ISBN 978-3-319-06519-9
13. Nimmagadda, S.L., Dreher, H.V.: On new emerging concepts of Petroleum Digital Ecosystem (PDE). J. WIREs Data Mining Knowl. Dis. **2**, 457–475 (2012). https://doi.org/10.1002/widm.1070
14. Nimmagadda, S.L.: Data Warehousing for Mining of Heterogeneous and Multidimensional Data Sources, Verlag Publisher, Scholar Press, OmniScriptum GMBH & CO. KG, Germany, pp. 1–657 (2015)
15. Parasnis, D.S.: Principles of Applied Geophysics. Chapman & Hall, USA (1997)

16. Vaishnavi, V., Kuechler, Jr., W.: Design Science Research Methods and Patterns: Innovating Information and Communication Technology, Auerbach Publications, NY, Taylor & Francis Group, Boca Raton, FL (2007)
17. Wight, A.W.R., Hare, L.H., Reynolds, J.R.: A Sedimentary Basin, NE Kalimantan, Indonesia: a century of exploration and future potential, Geological Society of Malaysia, Circum – Pacific Council for Energy and Mineral Resources (1992)

On a Smart Digital Gender Ecosystem for Managing the Socio-Economic Development in the African Contexts

Christina Namugenyi[1], Shastri L. Nimmagadda[2(✉)],
and Neel Mani[3(✉)]

[1] School of Information Technology, Monash University,
Johannesburg, South Africa
[2] School of Management, CBS, Curtin University, Perth, WA, Australia
shastri.nimmagadda@curtin.edu.au
[3] School of Computer Science, University College Dublin,
Dublin, Ireland
neelmanidas@gmail.com

Abstract. The research is aimed at investigating gender equality that affects the social and economic development in African countries. The existing data sources are examined based on the research question: "What is the role of female education and workforce on the social and economic development in African contexts, especially in East Africa?" To explore this question, we investigate the socio-economic development indicators such as employment, education, and population growth. We examine the effects of female population on impending education and employment indicators. For analysing the benchmarks, the empirical and observational research methods are deployed. Various data schemas are designed to test the gender based ecosystems and their models. Our results show the data relationship has a very strong positive connection between *female education* and *social economic development* dimensions. This is validated by qualitative research, obtained from questionnaires from one of the municipalities. We compute the polynomial regressions based on data fluctuations in large size gender ecosystems. We provide new regression models for visualization and knowledge interpretation. We conclude that *female education* and *employment* play characteristic roles in the social and economic development. Orthogonal polynomial models provide a scope of understanding the statistical inferences on *equality in education, quality of life* and other socio-economic insights that contribute to making knowledge-based solutions smartly and decisively.

Keywords: Gender · Digital ecosystem · African contexts
Socio-economic indicators

1 Introduction

The population of African continent is about 1.2 billion, comparable to the population size of China. It opens a large gender space in Africa. In gender perspective, though male population has an overall advantage, the size of female population evokes

© Springer International Publishing AG, part of Springer Nature 2019
I. Czarnowski et al. (Eds.): KES-IDT 2018, SIST 97, pp. 139–149, 2019.
https://doi.org/10.1007/978-3-319-92028-3_14

dominance in education, and participation in community development services. Monitoring socio-economic development is not possible without measuring the progress, creating an opportunity for efficient use of human, education, financial resources at local council and district levels. The existing data sources and statistics are required for designing, planning, implementing, monitoring and evaluating future forecasts of resources in geographically distributed African gender ecosystems. Relevant data and information need to be disseminated among various stakeholders for positive and effective-informed planning and management, knowledge based decisions and service deliveries.

Education is one of the key strategies in demand for improving the social and economic development [7] in Africa. However, there are constraints in reaching the general population and cultivating the message of education in many African regions. Besides, the continent faces an unequal distribution of education facilities between male and female population, envisaged as one of the major obstacles of economic development in different societies [6]. This is common in less economically developed societies in Africa, where female population is excluded from education systems (despite the progress in recent years) thus hindering gender development. In Uganda, it is estimated 31 million primary school girls including lower secondary school girls were out of school in 2013 [9]. In many African contexts, lack of data, information and knowledge on socio-economic indicators are hurdles that indeed motivate us to initiate the study.

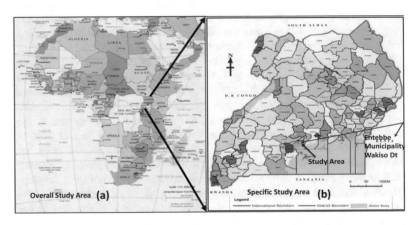

Fig. 1. Map of African continent showing countries, their territories and ethnicities and (b) focused study area for qualitative data analysis (Data sources: [6, 7, 9])

East Africa is one of the many regions whose economic development is constrained with lack of education and women empowerment. The statistics [6, 7] suggest large number of school drop outs in East Africa. The need for laborers to work in large plantations is very high, hence forcing farmers to resort to using their female children as a source of cheap labor. In broader context of Africa, the gender ecosystem needs to strengthen female education and women empowerment, counteracting the slow progress of socio-economic development. As illustrated in Fig. 1, we identify and ascribe several data entities and dimensions in digital map views, interpreting socio-economic indicators for populated countries in both periodic and geographic dimensions. Economic development of a country depends on its natural resources, their judicial use and how best we conserve them for future generations. The social and economic development indicators vary among different countries and districts depending on the rate at which female education and labor participation have been prioritized.

2   Issues and Challenges

In many countries and districts in Africa, there is an unequal distribution of female education in different parts of the continent especially at secondary and tertiary education levels. However, among many sub counties of the districts and municipality levels, a rapid change is observed as a result of encouragement in female education [7]. The research investigates for diverse ways to approach a solution broadly by analyzing the resources, cross cultural issues, education, healthcare, production, community building, works and planning and finance management entities. To support an evidence based policy and monitor the socio-economic development, grassroots planning, human resources, such as children education, gender issues, healthcare for all, community development, and poverty alleviation need great attention. Besides, lack of data, information and detailed knowledge of resource distribution preclude designing and developing new tools and technologies for data science of socio-economic yardsticks. We intend to build dimensional data models including a warehouse repository [2, 3] for documenting and organizing large size multidimensional data and information. For the purpose of capacity building and planning, without compromising the quality of life, peace and stability, the existing resources must be known in advance and used for testing the data models, before alternative solutions are offered. Various data sources [6, 7, 9], which are in the order of several gigabytes are used in the modelling studies.

3   Description of a Digital Gender Ecosystem

The gender is characterized by an ecosystem that differentiates female from male entity, to describe human beings with general behavior, social interactions, and fundamental rights with sense of self [9], all interpreted in various data entities and dimensions. The ecosystem implies that the gender is a complex set of relationships among various dimensions including their connectivity. The socio-economic indicators vary in size both periodically and geographically. In data management perspective, the multiple dimensions are used to build logical constructs and physical schemas for meta-metadata descriptions of gender ecosystem. With the result, if any part of the ecosystem changes, the other measures do change including their connected semantic, schematic and syntactic contents [8] and contexts. When a gender ecosystem generates a meaningful information with an interpretable and evaluable new knowledge, we say that the ecosystem may be sustainable [4] to exist for longer periods and at varying geographies. It is important to conceptualize the gender as an integral part of the ecosystem. We can use and reuse the integrated metadata models, besides testing of gender relationships in both periodic and geographic contextual attribute dimensions.

4 Research Objectives

1. To acquire data from existing sources; to identify attributes for building various data models, representing the education, socio-economic and employment indicators.
2. To make connections between data attribute dimensions through description of domain ontologies and integrate in multidimensional warehouse repositories for meta-metadata. Extract data views from metadata for visualization and interpretation.
3. To validate and analyze the metadata qualitatively and quantitatively and explore new knowledge and forecast future resources through data mining models.

5 Methodology

We cover the first two research objectives in this section. Multidimensional data modelling, data warehousing and mining are adopted for logical data organization and documentation in warehouse repositories. A questionnaire is designed to acquire the data including opinions and views of various schools in Uganda in particular in Wakiso district context. Both qualitative and quantitative methods [5] are used for interpreting the data acquired from various education institutions to deduce patterns, correlations and trends of various data attribute dimensions and their fact instances. As given in Tables 1, 2 and 3, several data dimensions and facts are interpreted. We document in Table 4 to document all the dimensions and facts of the data in repositories and make connections among facts of *female populations*, *socio-economic indicators* and *employment status*.

Table 1. Attribute dimensions and typical instances

Dimension	Number of data attributes	Number of Instances
Year	3	174
Female Work Force (ILO)	5	290
Country	5	290
Female to Male Labour Force Ratio	6	348
School Enrolment, Secondary	4	232
Population, Female	5	290
Population, Male	6	348
Gender Ratio	2	116
Population Density	5	290
Mortality Rate, Female	6	348
Population, Age Group	5	290
Mortality Rate, Infants	3	174
Pupil-Teacher Ratio, Secondary	5	290
Pupil-Teacher Ratio, Primary	6	348
Sec Education, Teachers Female	4	232
Primary Education, Teachers Female	6	348
Sex Ratio, at Birth	2	116

5.1   Description of Data Attributes and Their Characteristics

Though we focus on the entire African continent, but the existing data sources [6, 7] are examined for 12 African countries with various data patterns and trends of socio-economic indicators. Female population is grouped in four working categories according to age 20–24, 25–29, 30–34 and 35–39. In our modelling and analysis, different bubble plot views are presented based on geographic regions and their ethnicity. The periodic "Year" and geographic "Country" are key attribute dimensions in all our modelling studies. Various other attributes and their instances documented are shown in Tables 1, 2 and 3. The attribute data dimensions at places are contextualized based on demography.

Table 2. Female population & Education indicators

Attributes	Data Instances
Working age group	Census (2002) 32, 480
Primary School Population, aged 6–12 years	12, 320
Secondary School Population aged 13–19 years	6340
Population aged 60 years and above	1,145
Sex Ratio of total population (census 2002)	97 males for 100 females
Population Density	980 persons/km^2
Infant mortality	76 deaths per 1000 live births
Under 5 years' mortality	138 deaths per 1000 live births
Pupil Teacher Ratio, (Primary, 2004)	1:48
Student Teacher Ratio (secondary, 2004)	1:19

Table 3. Education and Economic Indicators and their Instances

Data Attributes	Number
Name of respondent	140
Number of places	140
Dates of responses	July – September 2017
Number of schools	100
Type of schools	5
Level of education	4
Age	>18
Gender	3
Employment	3 levels
Income	4 levels
Number of years employed	5 levels
Marital status	4 levels
Children level of education	3 levels
Literacy rate	66.8%
Gross primary enrolment ratio	52%

Table 4. Dimension and fact table used in modelling (for connectivity)

Dimensions	Attributes & Types	Facts	Connectivity
Population	Popu_ID, Num	(FPEI_Fact_ID); SOCI_Fact_ID	Yes
Name of Person	Name_ID Numb	(FPEI_Fact_ID); ECOI_Fact_ID	Yes
Place	Place_ID; Alpha	(FPEI_Fact_ID); EMPI_Fact_ID	Yes
Period	Period_ID; Date	(FPEI_Fact_ID)	Yes
School	School_ID; Alph	(FPEI_Fact_ID)	Yes
Type School	Type_ID; Alphabetic	(FPEI_Fact_ID)	Yes
Level of Education	Level_ID; Alphabetic	(FPEI_Fact_ID)	Yes
Gender	Gender_ID; Alpha	(FPEI_Fact_ID)	Yes
Enrolment	Enrol_ID; Number	(FPEI_Fact_ID)	Yes
Country	Country_ID, Alphabetic	(FPEI_Fact_ID)	Yes
Age	Age_ID, Number	(SOCI_Fact_ID)	Yes
Marriage	Marital_ID; Alpha	(SOCI_Fact_ID)	Yes
Status	SocStat_ID; Alpha	(SOCI_Fact_ID)	Yes
Soc Stat Type	SocStatType_ID; Alphabetic	(SOCI_Fact_ID); FPEI_Fact_ID	Yes
PerCapitaProduct	PerCapProd_ID; Alphabetic	ECOI_Fact_ID; FPEI_Fact_ID	Yes
PerCapitaPerDay	PerCapita_ID; Alphabetic	ECOI_Fact_ID	Yes
Econo	Econo_ID; Alphabetic	ECOI_Fact_ID	Yes
Earnings	Earning_ID; Alphabetic	EMPI_Fact_ID; FPEI_Fact_ID	Yes
Employ	Employ_ID; Alphabetic	EMPI_Fact_ID;	Yes
Years Employed	Year_ID; Numeric	EMPI_Fact_ID;	Yes

5.2   Modelling the Data Attribute Dimensions and Facts

Various data entities are identified from socio-economic indicators from which number of attributes and their instances are interpreted and documented for modelling [6, 7]. As shown in Fig. 2a, a data schema or model is articulated with multiple attribute dimensions and their fact instances in the study area.

Figure 2b demonstrates a conceptual ecosystem framework, developed for analyzing the data acquired in different parts of the study area. The framework brings together the data from human, socio-economic and employment ecosystems for integrating contextualized attribute dimensions and arriving at a warehoused metadata or repository.

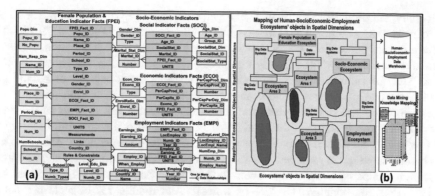

Fig. 2. Data model depicting the attribute dimensions, (b) a conceptual framework drawn for human-socio economic-employment ecosystems

6 Analysis and Discussions

The research objective 3 (in Sect. 4) covers the evaluation and implementation of the models in the study area. The existing data are used for building data schemas and implementing their metadata cubes through visualization and interpretation. Several data mining operations [1, 8] are carried out on cuboid data structures. The data views are interpreted for any anomalous data features and their visualizations in metadata. Using existing female and male populations, gender ratios are extracted for 12 African countries. As shown in Figs. 2 and 3, a phenomenal increase of population is interpreted in all female age groups, but the age group 20–24 is more anatomical in between years 1990–2017 (Fig. 3a). Year 1970 suggests low population rate in all age groups in spite of steady population densities, observed for all 12 countries as shown in Fig. 3b.

The data views presented in various bubble plots show multiple dimensions in different bubble sizes, densities and trends (orientations). As shown in Fig. 3a, the gender ratio instance above "1" is interpreted to have a female population dominance in many African countries. Overall interpretation of bubble plot views suggests that East and Southern African countries show phenomenal increase in female population with periodic dimension "Year". Nigeria in West Africa and Uganda in East Africa appear to a have rapid rate of population densities with same level of increase in female population compared with male population. Female work participation is another key dimension in our research and as shown in Figs. 4a and b, the bubble plot views show more or less a steady trend of female work participation in all 12 countries. Bubble plots of female work participation too exhibit interesting trends for East Africa and West Africa. An increase in the female employment in East African contexts is prominent, facilitating better socio-economic development programs in these regions.

Education and school enrolment in primary and secondary are other key dimensions of gender ecosystem research (Figs. 5a and b). To further validate and authenticate the bubble plots presented in Figs. 2, 3, 4 and 5, we have done a questionnaire survey in Ugandan contexts. We choose the Entebbe (of Wakiso District) municipality as a digital gender ecosystem. In Uganda in 2015, 100,831 boys were enrolled into university, but only 78,738 girls were enrolled showing a gap difference of 22,093. As per census, the latest population in a county of urban center of Entebbe is 79,700. The current female population is 27,951 with 50.8% more than male population. Expected pregnant females are 4000. Literacy rate of female population for age group 15–24 is 92%. The questionnaire data are qualitatively analyzed with support of bubble plot results. As one of the key socio-economic dimensions, the education attainment is analyzed for persons aged 15 years and above, since by that age chances are high that one cannot enroll in school if they had not. 10% had no formal education, followed by 47% who had completed primary education, 33.4% attained secondary education and 9% had attended tertiary institutions. Ten percent of the male population had no formal education compared to 24% of women. 29% of male population had completed secondary education compared to 22% of women. And finally, 7% of male population attended higher education as compared to 5% of women. As validated in a 2D bubble plot in Fig. 6, female population in between age range 10–22 attains more education than male population.

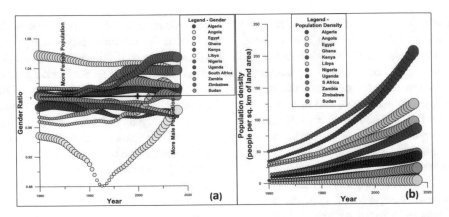

Fig. 3. Bubble plot data views representing (a) gender ratios and (b) population density

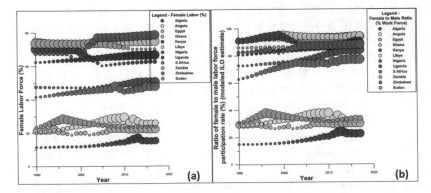

Fig. 4. Bubble plot views drawn for (a) Female work participation and (b) Ratio of female to male work participation attributes

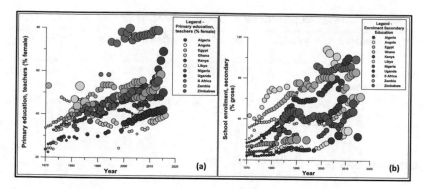

Fig. 5. Bubble plot views of enrollment in (a) primary and (b) secondary education

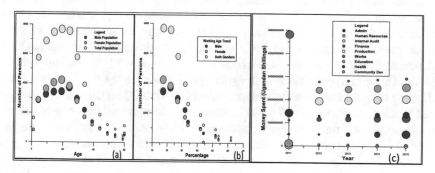

Fig. 6. Educated and working population size in an age attribute in the study area dimension

As shown in Fig, 6a, the educated populated age range in the study area is 10–30 years and this trend corroborates with working age group as shown in Fig. 6b. Similarly, the resources used in various administrative, human resources, internal audit, finance, production, public works and planning, education, health and community development entities are plotted in Fig. 6c. Though a small size of questionnaire data are analysed in the study area, the overall trend shows high population with female work and education indicators. For curvilinear trends, the polynomials are appropriate [4, 8] for fitting the observed data. The orthogonal polynomial regression is appropriate and at times necessary for higher order polynomial fits, if we need to explore the deeper knowledge of the gender ecosystem. The orthogonal polynomial factors have been converted to polynomial regression equations (with coefficients) from which Y is calculated from X. We compute orthogonal polynomials for three different socio-economic indicators as given in Tables 4 and 5 for smarter future forecasts of human resources in couple of countries. These models are further validated as shown in Figs. 3, 4, 5 and 6.

Table 5. Orthogonal polynomial fit between dimensions of female work force

Country	Orthogonal Polynomial Fit between dimensions of Female Work Force	Dimensions
Algeria	Y = 22553.79252 − 22.685345 * X + 0.005707283772 * pow(X,2)	Year, Female Work
Angola	Y = 9263.890925 − 9.197955787 * X + 0.002301705952 * pow(X,2)	Year, Female Work

7   Conclusions and Future Work

The methodology is robust in documenting, modelling and analyzing the attribute dimensions and fact instances of data, so as to make connections in metadata among female populations and other economic indicator attributes. Education and employment status facts are used for building multidimensional warehouse repositories. The existing data sources for 12 African countries are analyzed qualitatively and quantitatively. Questionnaire and secondary data collection methods are comparable in the qualitative and quantitative analysis, for case observed for sub county of Wakiso district of Uganda. There is an opportunity of integrating and analyzing the data from all districts of Uganda to make a comprehensive meta-metadata model. Metadata can facilitate the social researchers, agencies and governmental and non-governmental organizations to make assessments, future forecasts and optimal use of resources. The orthogonal polynomial models computed in between various attribute dimensions of the gender ecosystem are useful and need further research that can establish an efficacy and validity of the polynomial regression approach in large gender ecosystems. With the increase in the attainment of education among the female population in recent years, the African continent has seen an improvement in its social and economic development.

References

1. Berson, A., Smith, J.S.: Data Warehousing, Data Mining & OLAP, pp. 205–219, 221–513. Mc Graw – Hill Education (India) Pty Ltd. (2004)
2. Coronel, C., Morris, S., Rob, P.: Database Systems, Design, Implementation and Management, Course Technology, Cengage Learning, USA (2011). Entebbe Municipal Council Statistical Report, Uganda (2012)
3. Nimmagadda, S.L., Dreher, H.: Design of Petroleum Company's Metadata and an Effective Knowledge Mapping Methodology. IASTED, Cambridge (2007)
4. Nimmagadda, S.L., Zhu, D., Rudra, A.: Knowledge base smarter articulations for the open directory project in a sustainable digital ecosystem. In: WWW 2017, Perth, Australia, ACM (2017). http://dx.doi.org/10.1145/3041021.3054769. ISBN 978-1-4503-7/17/04
5. Research Methods Knowledge Base (2002), http://trochim.human.cornell.edu/kb/index.htm. (1 of 2). 21 July 2002 1:29:54. http://www.anatomyfacts.com/research/researchmethods knowledgebase.pdf
6. The World Bank (2016). https://data.worldbank.org/data-catalog/world-development-indicators

7. PRB Inform Empower Advance (2003). http://www.prb.org/Publications/Reports/2003/EmpoweringWomenDevelopingSocietyFemaleEducationintheMiddleEastandNorthAfrica.aspx
8. Nimmagadda, S.L.: Data Warehousing for Mining of Heterogeneous and Multidimensional Data Sources, Verlag Publisher, Scholar Press, OmniScriptum GMBH & CO. KG, Germany, pp. 1–657 (2015)
9. UNICEF (2013). https://www.unicef.org/publications/index_73682.html

A Framework for a Dynamic Inter-connection of Collaborating Agents with Multi-layered Application Abstraction Based on a Software-Bus System

Robert Brehm[1]([✉]), Mareike Redder[2], Gordon Flaegel[2], Jendrik Menz[3], and Cecil Bruce-Boye[2]

[1] University of Southern Denmark, Sonderburg, Denmark
`brehm@mci.sdu.dk`
[2] University of Applied Sciences Luebeck, Luebeck, Germany
[3] cbb Software GmbH, Luebeck, Germany

Abstract. With the currently ongoing and necessary process to define standards for multi-agent systems, their potential for adaptation to specific underlying system requirements becomes increasingly challenging. Especially in those applications, where multi-agent systems are coupled to hard- and software systems, which have specific operational characteristics, such as specific response times or quality-of-service requirements. As a result, it is proposed to separate the high-level decision making based on standardized multi-agent systems and the low-level system control into two layers. In this paper, based on a publisher/subscriber software-bus system, we propose a framework which allows dynamic allocation and linking of agents to underlying low-level hardware control. The concept of the framework and its architecture is introduced and an application example is outlined.

Keywords: Multi-agent system framework · Middleware architecture
OPC-SA

1 Introduction

The objective of cyber-physical systems (CPS) is to break traditional rigid centralized approaches and transform these into networks of interacting computational and physical devices. Therefore, system infrastructures are in a transition to a heterogeneous combination of hardware and software components. The communication infrastructure needs to be capable of processing various interfaces for different protocols as shown in [4,9].

CPS use high level decision making capabilities which involve autonomic decision making for cooperative systems and collaborative decision making based on

© Springer International Publishing AG, part of Springer Nature 2019
I. Czarnowski et al. (Eds.): KES-IDT 2018, SIST 97, pp. 150–157, 2019.
https://doi.org/10.1007/978-3-319-92028-3_15

negotiation [7]. These capabilities can be implemented using multi-agent systems (MAS). As such, the goal of CPS is to partition a complex problem that is traditionally solved in a centralized structure, into a decentralized network of modular, intelligent, adaptive and plug-able components. The overall control behaviour emerges as result of the interaction of the individual components. Important characteristics of CPS are modularity, scalability and thereupon flexibility. Autonomous decentralized agents ensure a distribution of objectives, decision processes, functionality and responsibilities in automation systems. In order to realize MAS, there are several approaches to define and standardize the agent communication such as the Agent Communication Language (ACL). These standards intend to make agent technology usable on a broad range of applications and thereby form the basis for the commercial deployment of the agent technologies.

Reconfigurability and real-time constraints are emerging aspects in industrial automation applications to provide low latency response to changing application demands and market conditions. Having CPS implemented for scalable and changing production systems, an agile handling of real-time operations requires low latency response and quality-of-service guarantee. Furthermore, a CPS enables computational and physical components to be linked and supports intelligent interaction. Therefore, the used technologies have to provide interfaces to various hardware and software elements. Commonly used technologies on the hardware control level in manufacturing processes are programmable logical controllers (PLC) or similar. Implementation of a decentralized MAS on those PLCs, which guarantees a certain quality-of-service, is not feasible. Hence, the intelligent decision making processes are carried out on a higher control level. Figure 1 which is adapted from [6,8] illustrates a high-level inter-agent communication connected to the hardware control level represented by a PLC. Figure 1 presents three communication channels which are used to operate and interact within a decentralized MAS system: 1. The high level inter-agent communication, 2. Low level real-time communication, 3. Communication between low and high level.

The inter-communication interface between the high-level agents and the lower hardware control layer is currently not standardized. The number of frameworks and platforms to implement MAS is increasing, a comprehensive overview and surveys are given by [1,5,12].

In this paper, a MAS framework based on a publisher/subscriber middleware architecture is presented. The focus for communication in MAS is on the various providers and standards of sensor technology and actuators for physical systems within the MAS, as agents and their decisions have a direct impact on their environment. The main advantage of an architecture which is based on publisher/subscriber middleware, is the abstraction of the low-level hardware layer, from a higher service oriented application (multi-agent decision making) layer. Further the publisher/subscriber middleware provides scalability, modularity, real-time capabilities and openness [14]. The proposed framework for agent communication is based on the Open Process Communication Simplified Architecture (OPC-SA) [13], which operates as a software bus system (recently renamed, former name was LabMap). The middleware acts as a hardware/software independent layer as shown in Fig. 1. The service oriented application communicates with the hardware through a set of strongly typified variables (registers) and OPC-SA carries out the actual transfer to the hardware through appropriate protocols and interfaces [3].

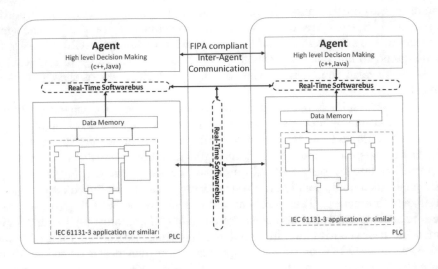

Fig. 1. Modified agent communication based on a software-bus, adapted from [6,8]

Hereafter in Sect. 2 we present a framework based on the software bus system OPC-SA, which enables an inter-agent communication with interoperability to the underlying hardware infrastructure. To demonstrate feasibility and necessity of the software-bus framework, in Sect. 3 a use case is introduced. Finally, we give concluding remarks and an outlook in Sect. 4.

2   Concept of the Framework

In CPS, the basic system in the physical layer has specific demands on the logic level above it, such as specific short and/or deterministic reaction times. Therefore the information processing is often implemented by a special hard- and software system, such as a Real-time Control System (RCS), for example PLCs. Part of the RCS is a communication infrastructure, for communication with peripheral system components and bus-connected RCSs. In order to ensure simple integration and preserve the quality-of-service (QoS) for the system and the infrastructure it relies on, it is a reasonable approach to embed agents in higher soft- and hardware layers. They provide the required abstraction based on high level inter-agent interaction in the MAS [7]. However MAS, based on non-deterministic communication protocols, such as ACL, cannot guarantee QoS as required by the RCS.

A characteristic of MAS is the physical and logical distribution of data and hardware. A software bus system which stretches through the MAS can provide distribution transparency for agents, which are connected to the software bus. A software bus architecture can be used as an underlying framework for MAS systems. Agents that produce information, create data and events and forward these to the software bus. Agents which consume data, create subscriptions with the software bus, which will notify them as subscribed data becomes available.

The herein used software bus middleware technology, OPC-SA provides distribution transparency and abstraction of the application layer from local and distributed data and hardware access. Thus, software components, such as agents, which are connected to the software bus system can react (and act) on remote data and hardware, in same way as on local data. The data access is subdivided into three I/O strategies: 1. a request only strategy in which the application receives a new data value after an explicit request, 2. data access via periodic polling, 3. obtain a new value via an event based strategy, every time a corresponding hardware value is about to change. The connection between agents is represented as publisher/subscriber relations. In OPC-SA, data and hardware are represented by devices of different classes, similar to classes in object orientated programming. The actual data can be accessed through registers, which provide the interface to devices. Registers are given a data type, initial value, time-stamp and an input/output direction to abstract information and the ability to synchronize any execution.

These OPC-SA abstractions for interfaces via devices provide inter-node communication via various communication protocols, such as e.g. CAN, EtherCAT, Modbus, TCP/IP over Ethernet and others. OPC-SA enables a notification service to the subscriber. Those services are subdivided into asynchronous and synchronous functions like callbacks, as asynchronous service or event based notifications to synchronize executions. This relation also allows, for example, a request for data does not have to take place directly from agent to agent, but can be executed across several nodes in a publisher/subscriber. Thus, the middleware provides a transparent data transfer within the publisher/subscriber chain without individual agents knowing more than their explicit neighbours. But executions are directly rooted via the middleware based relations from source to destination and can be changed at any time [3].

The inner architecture of the proposed framework is shown in Fig. 2. The core of the framework is a meet-and-greet mechanism, which is used for dynamic binding of collaborating agents into a MAS, this is indicated by the MEET and GREET blocks in Fig. 2. The affiliation of an agent can be configured by a configuration file. This specifies the local link topology of an agent for communication and to which corresponding agents, or type of agents, the agent belongs. Further it defines which information is to be exchanged with affiliated agents. Meaning, the MAS an agent belongs to, and the data which should be exchanged in the MAS with other agents. On this basis the framework provides dynamic binding and distributed data access between agents within a MAS.

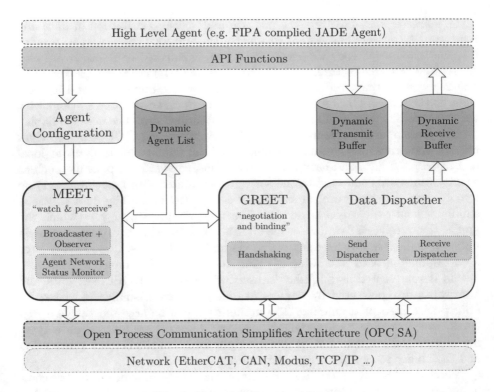

Fig. 2. Meet-and-Greet architecture

The MEET block implements the agents recognition mechanism to find corresponding agents. It invokes a periodic broadcast request message, which is sent repeatedly, at a specific interval to all agents in the network. An observer monitors the respond of active agents in the network and compares the result to the configuration (link topology, affiliated agents) of the local agents. If an affiliated agent is found, it is added to a dynamic agent list. Further, the GREET block is informed, which starts the agents negotiation and binding mechanism. The available corresponding agent is contacted directly, and both agents exchange information about the connection details and create the corresponding data OPC-SA registers for data exchange as defined in the configuration file. All information about corresponding agents are stored within the dynamic agent list. The communication between agents is abstracted through dynamic send and receive buffering, as indicated in Fig. 2. The overlaying application can add data for transmission to the dynamic transmit buffer, through an API function. Further, the overlaying application can receive data through a receive interface, which provides access to the dynamic receive buffer. Data from both, the dynamic transmit buffer and the dynamic receive buffer is dispatched by a data dispatch block. Part of the observer in the MEET block is a monitoring mechanism which detects dead and non-responding agents in the network. This monitoring mechanism senses the absence of an agent, by continuously comparing the respond to the broadcast message in the MEET mechanism, with the entries in the dynamic agent list. An agent which has been added with an active status to the dynamic agent list and does not response to the broadcast message anymore, will be removed from the list of active agents. In the following section a use case example of the proposed framework is introduced.

3   Application Example

The electrical energy supply system topology is currently in transition from a centralized topology towards an architecture which is decentralized and based on fluctuating and intermittent renewable electrical energy resources. One solution for efficient integration of distributed renewable energy resources are virtual microgrids (VMG) [10,11]. It is imperative that storage capacities and shift-able loads are operated efficiently to guarantee the entire exploitation of all resources. Several research projects demonstrate the use of decentralized cooperative multi-agent systems which incorporate a decentralized energy management approach to ensure flexibility, scalability and transparency. VMG are small-scale energy distribution grids, which cluster local energy resources, storage components and loads behind a common transformer substation. These VMG are operated with respect to specific operational objectives, such as minimization of weather dependent energy flow over a common transformer substation [2].

As a result of energy efficiency policies, electrical vehicles (eV) will successively replace fuel powered vehicles, this results in a growth of shiftable coincidentally unpredictable loads. The charging and discharging patterns of eV depend strongly on individual human behavior. In order to achieve the operational objectives within a micro grid, such as minimization of energy flow over a common transformer substation, a dispatch plan for storage capacities and eVs has to be compiled by the energy management system. Coordinated charging of eV can reduce the congestion and peak shaving for the utility grid. Implementing a multi-agent system in a micro grid that contains electrical vehicles requires the multi-agent system to react on changes with the environment (plug in or unplug of eV). Moreover a dynamic binding of the agent interconnect had to be provided. Therefore the used framework needs to provide an automatic establishment and destruction of inter-connections.

The process of coordinated charging involves the low level battery management system of the electrical vehicle and the charge station the electrical vehicle is connected to. The vehicle requests a specific amount of energy, to refill the battery, from the charge station. A low-level protocol, which is based on pulse width modulation (PWM) is used for communication between charge station and eV. The PWM duty cycle is used to communicate the maximum available charge power to the vehicle internal charge controller. Further a bidirectional half-duplex CAN bus (Controller Area Network) is used, in specific charge modes, to exchange information about the state-of-charge of the battery and the maximum charge currents. Based on a given deadline for the charging process, an agent in the charge station coordinates a schedule with neighbouring charge stations, in order to charge all electrical cars such that the operational objectives of the micro-grid are full-filled, such as to minimize peak loads. The actual charging process however involves a low level charge controller inside the battery management system of the car which can involve a low level CAN-bus agent which communicates via CAN bus or PWM with the charge station. In this situation the high level agent inside the charge station needs to be able to dynamically bind to the battery management system inside the car, which is only accessible via CAN bus. Here the used framework is able to dynamically bind to the cars CAN bus or to agents which can be implemented inside the battery management system of the car and which are accessible via CAN bus.

4 Conclusion

The information processing in a mechatronic system is often implemented on special hardware to guarantee a specific quality-of-service to the operational software system controlling it. These hard- and software systems are often incapable of hosting and implementing full featured standardized agents. Therefore the emphasis in this work has been laid on a framework for interfacing of standardized agents with low level hard- and software, which is an open research question to be answered.

The concept of a framework based on the Open Process Communication Simplifies Architecture (OPC-SA) software bus has been introduced. It has been shown that the framework provides distribution transparency for remote data and hardware, such that agents which are connected to the software bus system can react (and act) on remote data (software) and hardware, in same way as on local data. Further, agents can dynamically bind to data and hardware based on their local preferences which are defined in a configuration file. The core concept of this architecture, which is based on a meet-and-greet scheme has been introduced in detail, and an illustrative use case is given to highlight the usefulness of the proposed framework.

References

1. Bordini, R.H., Braubach, L., Dastani, M., Seghrouchni, A.E.F., Gomez-Sanz, J.J., Leite, J., O'Hare, G., Pokahr, A., Ricci, A.: A survey of programming languages and platforms for multi-agent systems. Informatica **30**(1), 33–44 (2006)
2. Brehm, R., Mátéfi-Tempfli, S., Top, S.: Consensus based scheduling of storage capacities in a virtual microgrid. Adv. Smart Syst. Res. **6**(1), 13–22 (2017). ISSN 2050-8662
3. Bruce-Boye, C., Kazakov, D.: Quality of uni-and multicast services in a middleware. LabMap study case. In: Innovative Algorithms and Techniques in Automation, Industrial Electronics and Telecommunications, pp. 89–94 (2007)
4. Kim, M., Kim, S.: A scenario-based user-oriented integrated architecture for supporting interoperability among heterogeneous home network middlewares. Comput. Sci. Appl. ICCSA **2006**, 669–678 (2006)

5. Kravari, K., Bassiliades, N.: A survey of agent platforms. J. Artif. Soc. Soc. Simul. **18**(1), 11 (2015)
6. Leitão, P., Colombo, A.W., Karnouskos, S.: Industrial automation based on cyber-physical systems technologies: prototype implementations and challenges. Comput. Ind. **81**, 11–25 (2016)
7. Leitao, P., Karnouskos, S., Ribeiro, L., Lee, J., Strasser, T., Colombo, A.W.: Smart agents in industrial cyber-physical systems. Proc. IEEE **104**(5), 1086–1101 (2016)
8. Marik, V., McFarlane, D.: Industrial adoption of agent-based technologies. IEEE Intell. Syst. **20**(1), 27–35 (2005)
9. Moon, K.D., Lee, Y.H., Lee, C.E., Son, Y.S.: Design of a universal middleware bridge for device interoperability in heterogeneous home network middleware. IEEE Trans. Consum. Electron. **51**(1), 314–318 (2005)
10. Nasri, M., Ginn, H.L., Moallem, M.: Application of intelligent agent systems for real-time coordination of power converters (RCPC) in microgrids. In: 2014 IEEE Energy Conversion Congress and Exposition (ECCE), pp. 3942–3949. IEEE (2014)
11. Olivares, D.E., Mehrizi-Sani, A., Etemadi, A.H., Cañizares, C.A., Iravani, R., Kazerani, M., Hajimiragha, A.H., Gomis-Bellmunt, O., Saeedifard, M., Palma-Behnke, R., et al.: Trends in microgrid control. IEEE Trans. Smart Grid **5**(4), 1905–1919 (2014)
12. Sturm, A., Shehory, O.: Agent-oriented software engineering: revisiting the state of the art. In: Agent-Oriented Software Engineering, pp. 13–26. Springer, Heidelberg (2014)
13. VDI-Richtlinie: Middleware in der Automatisierungstechnik - Grundlagen, Richtlinien Nr. 2657, Blatt 1 (2013)
14. Weyns, D., Helleboogh, A., Holvoet, T., Schumacher, M.: The agent environment in multi-agent systems: a middleware perspective. Multiagent Grid Syst. **5**(1), 93–108 (2009)

Non-reciprocal Pairwise Comparisons and Solution Method in AHP

Kazutomo Nishizawa$^{(\boxtimes)}$

Nihon University, 1-2-1 Izumicho, Narashino, Chiba 275-8575, Japan
nishizawa.kazutomo@nihon-u.ac.jp

Abstract. In this study of Analytic Hierarchy Process (AHP), a modified paired comparison method and its solution method are proposed. In traditional AHP, pairwise comparisons were carried out using a nine-stage evaluation process, and then the pairwise comparison matrix was constructed by the decision maker. The elements of the pairwise comparison matrix have the property of reciprocity. For the decision maker, this property is not easy to judge for pairwise comparisons. Hence, we would like to reduce the burden of judgment for the decision maker. The purposes of this study are to construct a non-reciprocal pairwise comparison matrix and to obtain perfectly consistent weights. By applying the proposed methods to examples, the process of these methods is illustrated.

Keywords: AHP · Comparisons · Non-reciprocal · Consistency

1 Introduction

In this study, a modified paired comparison method and its solution method in Analytic Hierarchy Process (AHP)[5], are proposed. In traditional AHP, pairwise comparisons by the decision maker were carried out using a nine-stage evaluation process, that is 1 to 9 and 1/2 to 1/9. Then the $n \times n$ pairwise comparison matrix $\boldsymbol{A}$ was constructed. Assume that $\boldsymbol{A}$ is a complete matrix of comparisons. The elements of $\boldsymbol{A}$, that is a_{ij} $(i, j = 1$ to $n)$, have the property of reciprocity. For example, $a_{12} = 9$ and $a_{21} = 1/9$. Of course $a_{11} = a_{22} = 1$.

For the decision maker, this property is not easy to judge for pairwise comparisons. Hence, we would like to reduce the burden of judgment for the decision maker. The purposes of this study are to construct a non-reciprocal pairwise comparison matrix and to obtain perfectly consistent weights.

The idea for these methods is based on previous studies. One is the estimation of unknown or missing pairwise comparisons [2], and the other is the adjustment from inconsistent comparisons [1,3,4].

Based on a previous paper [4], from an inconsistent comparison matrix, we can obtain a perfectly consistent one, that is $\lambda_{max} = n$. Therefore we consider that there is no need to satisfy the property of being reciprocal.

© Springer International Publishing AG, part of Springer Nature 2019
I. Czarnowski et al. (Eds.): KES-IDT 2018, SIST 97, pp. 158–165, 2019.
https://doi.org/10.1007/978-3-319-92028-3_16

By applying the proposed methods to the examples, the process of these methods is illustrated.

This paper consists of the following sections. In Sect. 2, the proposed comparison method is explained. In Sect. 3, the proposed calculation method is shown. In Sect. 4, the application of the proposed methods to examples is illustrated and discussed. And finally, in Sect. 5, we discuss the conclusions of this study.

2 Proposed Comparison Method

In this section, the proposed comparison method is explained. The idea of this method was developed in a previous study on the estimation of unknown or missing pairwise comparisons [2].

A part of a proposed pairwise comparison sheet is shown in Fig. 1.

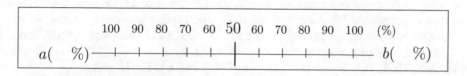

Fig. 1. A part of a pairwise comparison sheet

In this sheet, a comparison between alternatives a and b is carried out. The decision maker judges that alternative b was $x\%$ better than a. Then the decision maker fills out the result in this sheet.

An example of a comparison sheet between a and b is illustrated in Fig. 2. Assume that the decision maker decides $x = 70$.

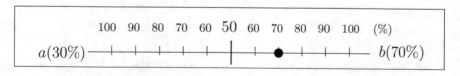

Fig. 2. An example of filling out a comparison sheet between a and b

Therefore the decision maker marks a dot at the position 70 on the b side, fills out the number 70 in the parentheses, and further fills out the number 30 on the a side.

After all comparisons are carried out, based on the comparison sheet, an $n \times n$ comparison matrix $\boldsymbol{A}$, a_{ij} ($i, j = 1$ to n), is constructed. In this example, the values of elements of $\boldsymbol{A}$ are $a_{ab} = 30$, $a_{ba} = 70$ and $a_{aa} = a_{bb} = 50$.

However, if the decision maker judges that b is 100% better than a, that is $x = 100$, then $a_{ab} = 0$ and $a_{ba} = 100$. In this case, $a_{ab} = 0$, is not good for calculating the eigenvector. So, we transform all comparisons as follows. If alternative i is $x\%$ better than j, then

$$a_{ij} = \theta^{(x/100)}, \; a_{ji} = \theta^{(100-x)/100}. \tag{1}$$

where θ is an arbitrary constant and $\theta > 1$, for example $\theta = 2$. In the case of $x = 100$, based on Eq. (1), we obtain $a_{ij} = \theta$ and $a_{ji} = 1$.

The order of comparisons in this proposed method are decided randomly.

3   Proposed Calculation Method

In this section, the proposed calculation method is shown. The idea of this method was developed in two previous studies; one is the estimation of unknown or missing pairwise comparisons [2] and the other is the adjustment from inconsistent comparisons [4]. Assume that, in this method, the number of alternatives is n, and the pairwise comparison matrix $\boldsymbol{A}$ consists of complete comparisons.

In the case of perfectly consistent comparisons, $\boldsymbol{A}$ has $\lambda_{max} = n$. For the elements of $\boldsymbol{A}$, that is a_{ij} ($i = 1$ to n and $j = 1$ to n), Eq. (2) holds:

$$a_{ij} = w_i/w_j, \tag{2}$$

where w_i is the eigenvector of alternative ai.

Based on Eq. (2), for any k ($k = 1$ to n), Eq. (3) is obtained.

$$a_{ij} = a_{kj}/a_{ki} \tag{3}$$

In the imperfectly consistent case, Eq. (2) does not hold. Therefore based on Eq. (3) and using the geometric mean, the first adjusted element $a_{ij}^{(1)}$ is obtained by Eq. (4).

$$a_{ij}^{(1)} = \sqrt[n]{\prod_{k=1}^{n} (a_{kj}/a_{ki})} \tag{4}$$

The second adjusted element $a_{ij}^{(2)}$ is calculated by $(a_{kj}^{(1)}/a_{ki}^{(1)})$, instead of (a_{kj}/a_{ki}), for $i = 1$ to n and $j = 1$ to n, in Eq. (4).

The $m - th$ adjusted element $a_{ij}^{(m)}$ is calculated by Eq. (5).

$$a_{ij}^{(m)} = \sqrt[n]{\prod_{k=1}^{n} (a_{kj}^{(m-1)}/a_{ki}^{(m-1)})} \tag{5}$$

If $|a_{ij}^{(m)} - a_{ij}^{(m-1)}| < \varepsilon$ is obtained, for example with $\varepsilon = 1.0^{-6}$, we can judge that the adjusted elements converge. Then the adjusted comparison matrix $\boldsymbol{A'}$ is obtained. As a result, $\boldsymbol{A'}$ has $\lambda_{max} = n$ and is perfectly consistent.

The procedure of the proposed calculation method, consisting of P1 to P4, is summarized as follows.

P1: Construct a non-reciprocal comparison matrix $\boldsymbol{A}$, using Eq. (1).
P2: Calculate the adjusted comparison elements using Eq. (5).
P3: Repeat Eq. (5) until $|a_{ij}^{(m)} - a_{ij}^{(m-1)}| < \varepsilon$.
P4: The eigenvector of $\boldsymbol{A'}$ is obtained from any column vector of $\boldsymbol{A'}$.

In this procedure, P3 gives us the perfectly consistent comparison matrix $\boldsymbol{A'}$. From the result of P4, the column vector of $\boldsymbol{A'}$ that has maximum value equal to 1, coincides with the result by the power method.

For the non-reciprocal comparison matrix, the perfectly consistent principle eigenvector is easily obtained by the proposed calculation method.

4 Examples

In this section, using the proposed calculation procedure, P1 to P4, the process for obtaining perfectly consistent weights from non-reciprocal comparison matrices, Example 1 and Example 2, are illustrated. These examples consist of four alternatives.

First, based on the proposed comparison method, the comparisons are carried out by the decision maker. Example 1 is one of the comparisons by the decision maker. We can construct the non-reciprocal comparison matrix using the proposed method. However, unfortunately, Example 1 does not include evaluation results of 100% or 0%. In actual cases, 100% evaluations are rare.

Example 2 is intentionally modified from Example 1 to include a 100% evaluation. Then we can obtain the perfectly consistent principal eigenvectors of Example 1 and Example 2 using the proposed methods.

4.1 Example 1

The comparison sheet of Example 1, judged by the decision maker, is shown in Fig. 3. This example, in Fig. 3, does not include evaluation results of 100% or 0%. Based on this comparison sheet, we can construct the comparison matrix in Eq. (6). This matrix is a non-reciprocal matrix.

$$\boldsymbol{A_1} = \begin{bmatrix} 50 & 75 & 90 & 80 \\ 25 & 50 & 90 & 85 \\ 10 & 10 & 50 & 20 \\ 20 & 15 & 80 & 50 \end{bmatrix} \tag{6}$$

According to proposed calculation method P1, using $\theta = 2$ in Eq. (1), the result $\boldsymbol{A_{1(\theta=2)}}$ is obtained in Eq. (7).

$$\boldsymbol{A_{1(\theta=2)}} = \begin{bmatrix} 1.414214 & 1.681793 & 1.866066 & 1.741101 \\ 1.189207 & 1.414214 & 1.866066 & 1.802501 \\ 1.071773 & 1.071773 & 1.414214 & 1.148698 \\ 1.148698 & 1.109569 & 1.741101 & 1.414214 \end{bmatrix} \tag{7}$$

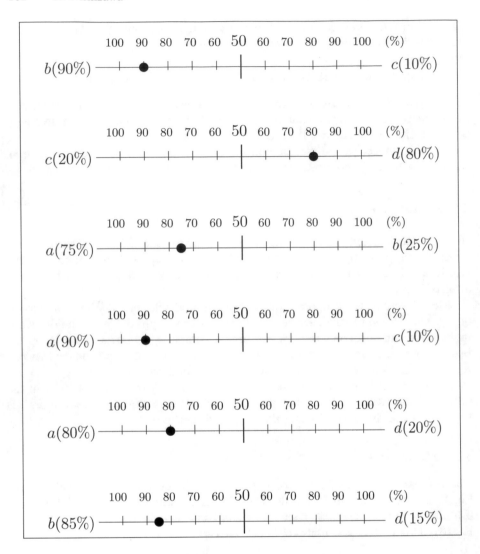

Fig. 3. An example of a pairwise comparison sheet

Next from P2, the perfectly consistent comparison matrix of Eq. (7) is calculated by Eq. (5). When the convergence condition in P3 is set to $\varepsilon = 1.0^{-6}$, we can obtain matrix $\boldsymbol{A}'_{1(\theta=2)}$ in Eq. (8).

$$\boldsymbol{A}'_{1(\theta=2)} = \begin{bmatrix} 1.000000 & 1.081100 & 1.426520 & 1.252664 \\ 0.924984 & 1.000000 & 1.319508 & 1.158694 \\ 0.701007 & 0.757858 & 1.000000 & 0.878126 \\ 0.798298 & 0.863040 & 1.138789 & 1.000000 \end{bmatrix} \tag{8}$$

From P4, the principal eigenvector of Eq. (8) is easily obtained in Eq. (9).

$$w_{1(\theta=2)} = \begin{bmatrix} 1.000000 \\ 0.924984 \\ 0.701007 \\ 0.798298 \end{bmatrix} \tag{9}$$

Since Eq. (6) does not include evaluation results of 100% or 0% so we can apply proposed Eqs. (5) to (6) directly. When $\varepsilon = 1.0^{-6}$ in P3, the perfect consistent matrix A_1' is obtained below.

$$A_1' = \begin{bmatrix} 1.000000 & 1.296278 & 4.053600 & 2.177939 \\ 0.771440 & 1.000000 & 3.127108 & 1.680148 \\ 0.246694 & 0.319784 & 1.000000 & 0.537285 \\ 0.459150 & 0.595186 & 1.861210 & 1.000000 \end{bmatrix} \tag{10}$$

From the first column of Eq. (10), the principal eigenvector is obtained in Eq. (11).

$$w_1 = \begin{bmatrix} 1.000000 \\ 0.771440 \\ 0.246694 \\ 0.459150 \end{bmatrix} \tag{11}$$

In this case, the order of alternatives in Eq. (11) is same as in Eq. (9), calculated by $\theta = 2$. However, the values of the principal eigenvector are different.

Therefore calculating with various values of θ, we can obtain a vector close to Eq. (11) when $\theta = 12$. Based on the proposed calculation method, using $\theta = 12$, Eqs. (12) and (13) are obtained below.

$$A_{1(\theta=12)} = \begin{bmatrix} 3.464102 & 6.447420 & 9.359726 & 7.300372 \\ 1.861210 & 3.464102 & 9.359726 & 8.266165 \\ 1.282089 & 1.282089 & 3.464102 & 1.643752 \\ 1.643752 & 1.451701 & 7.300372 & 3.464102 \end{bmatrix} \tag{12}$$

$$A_{1(\theta=12)}' = \begin{bmatrix} 1.000000 & 1.322537 & 3.573390 & 2.242508 \\ 0.756122 & 1.000000 & 2.701920 & 1.695610 \\ 0.279846 & 0.370107 & 1.000000 & 0.627557 \\ 0.445929 & 0.589758 & 1.593480 & 1.000000 \end{bmatrix} \tag{13}$$

From Eq. (13), as a result, the principal eigenvector, Eq. (14), is obtained.

$$w_{1(\theta=12)} = \begin{bmatrix} 1.000000 \\ 0.756122 \\ 0.279846 \\ 0.445929 \end{bmatrix} \tag{14}$$

In this case, Eq. (14) is close to Eq. (11), so $\theta = 12$ seems to be an appropriate value for Example 1.

4.2   Example 2

The next example, Example 2, is intentionally modified from Example 1 because 100% evaluations are rare cases. The modified matrix, A_2, is shown in Eq. (15).

$$A_2 = \begin{bmatrix} 50 & 75 & 100 & 80 \\ 25 & 50 & 90 & 85 \\ 0 & 10 & 50 & 20 \\ 20 & 15 & 80 & 50 \end{bmatrix} \tag{15}$$

Based on the proposed calculation method when $\theta = 2$, Eqs. (16) and (17) are obtained.

$$A_{2(\theta=2)} = \begin{bmatrix} 1.414214 & 1.681793 & 2.000000 & 1.741101 \\ 1.189207 & 1.414214 & 1.866066 & 1.802501 \\ 1.000000 & 1.071773 & 1.414214 & 1.148698 \\ 1.148698 & 1.109569 & 1.741101 & 1.414214 \end{bmatrix} \tag{16}$$

$$A'_{2(\theta=2)} = \begin{bmatrix} 1.000000 & 1.099997 & 1.476826 & 1.274561 \\ 0.909093 & 1.000000 & 1.342573 & 1.158694 \\ 0.677128 & 0.744839 & 1.000000 & 0.863040 \\ 0.784584 & 0.863040 & 1.158694 & 1.000000 \end{bmatrix} \tag{17}$$

From Eq. (17), as a result, the principal eigenvector is obtained below.

$$w_{2(\theta=2)} = \begin{bmatrix} 1.000000 \\ 0.909093 \\ 0.677128 \\ 0.784584 \end{bmatrix} \tag{18}$$

Furthermore when $\theta = 12$, Eqs. (19), (20) and (18) are obtained by the proposed calculation method.

$$A_{2(\theta=12)} = \begin{bmatrix} 3.464102 & 6.447420 & 12.000000 & 7.300372 \\ 1.861210 & 3.464102 & 9.359726 & 8.266165 \\ 1.000000 & 1.282089 & 3.464102 & 1.643752 \\ 1.643752 & 1.451701 & 7.300372 & 3.464102 \end{bmatrix} \tag{19}$$

$$A'_{2(\theta=12)} = \begin{bmatrix} 1.000000 & 1.407302 & 4.046126 & 2.386236 \\ 0.710579 & 1.000000 & 2.875094 & 1.695610 \\ 0.247150 & 0.347815 & 1.000000 & 0.589758 \\ 0.419070 & 0.589758 & 1.695610 & 1.000000 \end{bmatrix} \tag{20}$$

$$w_{2(\theta=12)} = \begin{bmatrix} 1.000000 \\ 0.710579 \\ 0.247150 \\ 0.419070 \end{bmatrix} \tag{21}$$

Also in Example 2 when $\theta = 12$, the obtained Eq. (21) is close to Eq. (11).

In Example 1 and Example 2, we consider $\theta = 12$ seems to be an appropriate value. However we not sure how to define an appropriate value of θ.

5   Conclusion

In this paper, a non-reciprocal pairwise comparison method in AHP and its solution method were proposed. These methods were based on a previous study to adjust inconsistent comparison matrices. Applying the proposed methods to two examples, the following was obtained.

1. Using the proposed comparison sheet, the pairwise comparisons could be easily carried out, and the burden of the decision maker was reduced.
2. From the non-reciprocal comparison matrix, the perfectly consistent matrix was obtained by the proposed calculation method.
3. We were able to easily obtain the principal eigenvector from the perfectly consistent matrix using the proposed method instead of the power method.

Furthermore the method proposed in this study requires verification with many different examples.

How to decide the most appropriate value of θ in Eq. (1) is a question for future research.

References

1. Nishizawa, K.: A consistency improving method in binary AHP. J. Oper. Res. Soc. Japan **38**, 21–33 (1995)
2. Nishizawa, K.: Estimation of unknown comparisons in incomplete AHP and it's compensation. In: Report of the Research Institute of Industrial Technology, Nihon University, vol. 77, pp. 1–8 (2005)
3. Nishizawa, K. : The improvement of pairwise comparison method of the alternatives in the AHP. In: Intelligent Decision Technologies. SIST, vol. 39, no. 1, pp. 483–491 (2015)
4. Nishizawa, K.: Adjustment from inconsistent comparisons in AHP to perfect consistency. In: Intelligent Decision Technologies. SIST, vol. 72, no. 1, pp. 293–300 (2017)
5. Saaty, T.L.: The Analytic Hierarchy Process. McGraw-Hill, New York (1980)

Super Pairwise Comparison Matrix
in the Multiple Dominant AHP
with Hierarchical Criteria

Takao Ohya[1]([⊠]) and Eizo Kinoshita[2]

[1] School of Science and Engineering, Kokushikan University, Tokyo, Japan
takaohya@kokushikan.ac.jp
[2] Faculty of Urban Science, Meijo University, Gifu, Japan
kinoshit@urban.meijo-u.ac.jp

Abstract. We have proposed a super pairwise comparison matrix (SPCM) to express all pairwise comparisons in the evaluation process of the dominant analytic hierarchy process (D-AHP) or the multiple dominant AHP (MDAHP) as a single pairwise comparison matrix and have shown calculations of super pairwise comparison matrix in MDAHP with hierarchical criteria. This paper shows calculations of super pairwise comparison matrix in MDAHP with hierarchical criteria.

Keywords: Super pairwise comparison matrix · The Multiple Dominant AHP
Logarithmic Least Square Method

1 Introduction

AHP (the Analytic Hierarchy Process) proposed by Saaty [1] enables objective decision making by top-down evaluation based on an overall aim.

In actual decision making, a decision maker often has a specific alternative (regulating alternative) in mind and makes an evaluation on the basis of the alternative. This was modeled in D-AHP (the dominant AHP), proposed by Kinoshita and Nakanishi [2].

If there are more than one regulating alternatives and the importance of each criterion is inconsistent, the overall evaluation value may differ for each regulating alternative. As a method of integrating the importance in such cases, CCM (the concurrent convergence method) was proposed. Kinoshita and Sekitani [3] showed the convergence of CCM.

Ohya and Kinoshita [4] proposed an SPCM (Super Pairwise Comparison Matrix) to express all pairwise comparisons in the evaluation process of the dominant analytic hierarchy process (AHP) or the multiple dominant AHP (MDAHP) as a single pairwise comparison matrix.

Ohya and Kinoshita [5] showed, by means of a numerical counterexample, that in MDAHP an evaluation value resulting from the application of the logarithmic least squares method (LLSM) to an SPCM does not necessarily coincide with that of the evaluation value resulting from the application of the geometric mean multiple D-AHP

© Springer International Publishing AG, part of Springer Nature 2019
I. Czarnowski et al. (Eds.): KES-IDT 2018, SIST 97, pp. 166–172, 2019.
https://doi.org/10.1007/978-3-319-92028-3_17

(GMMDAHP) to the evaluation value obtained from each pairwise comparison matrix by using the geometric mean method.

Ohya and Kinoshita [6] showed, using the error models, that in D-AHP an evaluation value resulting from the application of the logarithmic least squares method (LLSM) to an SPCM necessarily coincide with that of the evaluation value resulting obtained by using the geometric mean method to each pairwise comparison matrix.

Ohya and Kinoshita [7] showed the treatment of hierarchical criteria in D-AHP with super pairwise comparison matrix.

Ohya and Kinoshita [8] showed the example of using SPCM with the application of LLSM for calculation of MDAHP.

Ohya and Kinoshita [9] showed that the evaluation value resulting from the application of LLSM to an SPCM agrees with the evaluation value determined by the application of D-AHP to the evaluation value obtained from each pairwise comparison matrix by using the geometric mean.

This paper shows calculations of SPCM in MDAHP with hierarchical criteria.

2 D-AHP and SPCM

This section explains D-AHP, GMMDAHP and a SPCM to express the pairwise comparisons appearing in the evaluation processes of D-AHP and MDAHP as a single pairwise comparison matrix. Section 2.1 outlines D-AHP procedure and explicitly states pairwise comparisons, and Sect 2.2 explains the SPCM that expresses these pairwise comparisons as a single pairwise comparison matrix.

2.1 Evaluation in D-AHP

The true absolute importance of alternative $a(a = 1,\ldots,A)$ at criterion $c(c = 1,\ldots,C)$ is v_{ca}. The final purpose of the AHP is to obtain the relative value between alternatives of the overall evaluation value $v_a = \sum_{c=1}^{C} v_{ca}$ of alternative a. The procedure of D-AHP for obtaining an overall evaluation value is as follows:

D-AHP

Step 1: The relative importance $u_{ca} = \alpha_c v_{ca}$ (where α_c is a constant) of alternative a at criterion c is obtained by some kind of methods. In this paper, u_{ca} is obtained by applying the pairwise comparison method to alternatives at criterion c.

Step 2: Alternative d is the regulating alternative. The importance u_{ca} of alternative a at criterion c is normalized by the importance u_{ca} of the regulating alternative d, and $u_{ca}^d (= u_{ca}/u_{cd})$ is calculated.

Step 3: With the regulating alternative d as a representative alternative, the importance w_c^d of criterion c is obtained by applying the pairwise comparison method to criteria, where, w_c^d is normalized by $\sum_{c=1}^{C} w_c^d = 1$.

Step 4: From u_{ca}^d, w_c^d obtained at Steps 2 and 3, the overall evaluation value $t_a = \sum_{c=1}^C w_c^d u_{ca}^d$ of alternative a is obtained. By normalization at Steps 2 and 3, $u_d = 1$. Therefore, the overall evaluation value of regulating alternative d is normalized to 1

2.2 SPCM

The relative comparison values $r_{c'a'}^{ca}$ of importance v_{ca} of alternative a at criteria c as compared with the importance $v_{c'a'}$ of alternative a' in criterion c', are arranged in a $(CA \times CA)$ or $(AC \times AC)$ matrix. This is proposed as the SPCM $\mathbf{R} = \left(r_{c'a'}^{ca}\right) or \left(r_{a'c'}^{ac}\right)$.

In a $(CA \times CA)$ matrix, index of alternative changes first. In a $(CA \times CA)$ matrix, SPCM's $(A(c-1)+a, A(c'-1)+a')$th element is $r_{c'a'}^{ca}$.

In a $(AC \times AC)$ matrix, index of criteria changes first. In a $(AC \times AC)$ matrix, SPCM's $(C(a-1)+c, C(a'-1)+c')$th element is $r_{a'c'}^{ac}$.

In a SPCM, symmetric components have a reciprocal relationship as in pairwise comparison matrices. Diagonal elements are 1 and the following relationships are true:

If $r_{c'a'}^{ca}$ exists, then $r_{ca}^{c'a'}$ exists and

$$r_{ca}^{c'a'} = 1/r_{c'a'}^{ca} \tag{1}$$

$$r_{ca}^{ca} = 1 \tag{2}$$

Pairwise comparison at Step 1 of D-AHP consists of the relative comparison value $r_{ca'}^{ca}$ of importance v_{ca} of alternative a, compared with the importance $v_{ca'}$ of alternative a' at criterion c.

Pairwise comparison at Step 3 of D-AHP consists of the relative comparison value $r_{c'd}^{cd}$ of importance v_{cd} of alternative d at criterion c, compared with the importance $v_{c'd}$ of alternative d at criterion c', where the regulating alternative is d.

SPCM of D-AHP or MDAHP is an incomplete pairwise comparison matrix. Therefore, the LLSM based on an error model or an eigenvalue method such as the Harker method [10] or two-stage method is applicable to the calculation of evaluation values from an SPCM.

3 Numerical Example of Using SPCM for Calculation of MDAHP

Let us take as an example the hierarchy shown in Fig. 1. Three alternatives from 1 to 3 and seven criteria from I to VI, and S are assumed, where Alternative 1 and Alternative 2 are the regulating alternatives. Criteria IV to VI are grouped as Criterion S, where Criterion IV and Criterion V are the regulating criteria.

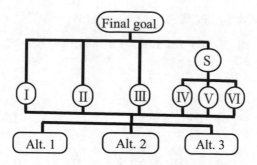

Fig. 1. The hieratical structure.

As the result of pairwise comparison between alternatives at criteria $c(c = I, \ldots, VI)$, the following pairwise comparison matrices $\mathbf{R}_c^A, c = I, \ldots, VI$ are obtained:

$$\mathbf{R}_I^A = \begin{pmatrix} 1 & \frac{1}{3} & 5 \\ 3 & 1 & 3 \\ \frac{1}{5} & \frac{1}{3} & 1 \end{pmatrix}, \mathbf{R}_{II}^A = \begin{pmatrix} 1 & 7 & 3 \\ \frac{1}{7} & 1 & \frac{1}{3} \\ \frac{1}{3} & 3 & 1 \end{pmatrix}, \mathbf{R}_{III}^A = \begin{pmatrix} 1 & \frac{1}{3} & \frac{1}{3} \\ 3 & 1 & \frac{1}{3} \\ 3 & 3 & 1 \end{pmatrix},$$

$$\mathbf{R}_{IV}^A = \begin{pmatrix} 1 & 3 & 5 \\ \frac{1}{3} & 1 & 1 \\ \frac{1}{5} & 1 & 1 \end{pmatrix}, \mathbf{R}_V^A = \begin{pmatrix} 1 & \frac{1}{3} & 3 \\ 3 & 1 & 5 \\ \frac{1}{3} & \frac{1}{5} & 1 \end{pmatrix}, \mathbf{R}_{VI}^A = \begin{pmatrix} 1 & \frac{1}{5} & 3 \\ 5 & 1 & 7 \\ \frac{1}{3} & \frac{1}{7} & 1 \end{pmatrix},$$

With regulating Alternative 1 and Alternative 2 as the representative alternatives, and Criterion IV and Criterion V as the representative criteria, importance between criteria was evaluated by pairwise comparison. As a result, the following pairwise comparison matrices $R_1^C, R_1^S, R_2^C, R_2^S$ are obtained:

$$R_1^C = \begin{bmatrix} 1 & \frac{1}{3} & 3 & \frac{1}{3} & \frac{1}{5} \\ 3 & 1 & 5 & 1 & \frac{1}{2} \\ \frac{1}{3} & \frac{1}{5} & 1 & \frac{1}{5} & \frac{1}{9} \\ 3 & 1 & 5 & 1 & \frac{1}{2} \\ 5 & 2 & 9 & 2 & 1 \end{bmatrix}, R_1^S = \begin{bmatrix} 1 & \frac{1}{2} & 2 \\ 2 & 1 & 5 \\ \frac{1}{2} & \frac{1}{5} & 1 \end{bmatrix},$$

$$R_2^C = \begin{bmatrix} 1 & 5 & 1 & 3 & \frac{1}{9} \\ \frac{1}{5} & 1 & \frac{1}{3} & 1 & \frac{1}{9} \\ 1 & 3 & 1 & 1 & \frac{1}{9} \\ \frac{1}{3} & 1 & 1 & 1 & \frac{1}{9} \\ 9 & 9 & 9 & 9 & 1 \end{bmatrix}, R_2^S = \begin{bmatrix} 1 & \frac{1}{9} & \frac{1}{4} \\ 9 & 1 & 6 \\ 4 & \frac{1}{6} & 1 \end{bmatrix},$$

The (CA × CA) order SPCM for this example is

	I1	I2	I3	II1	II2	II3	III1	III2	III3	IV1	IV2	IV3	V1	V2	V3	VI1	VI2	VI3
I1	1	1/3	5	1/3			3			1/3			1/5					
I2	3	1	3		5			1			3			1/9				
I3	1/5	1/3	1															
II1	3			1	7	3	5			1			1/2					
II2		1/5		1/7	1	1/3		1/3			1			1/9				
II3				1/3	3	1												
III1	1/3			1/5			1	1/3	1/3	1/5			1/9					
III2		1			3		3	1	1/3		1			1/9				
III3							3	3	1									
IV1	3			1			5			1	3	5	1/2			2		
IV2		1/3			1			1		1/3	1	1		1/9			1/4	
IV3										1/5	1	1						
V1	5			2			9			2			1	1/3	3	5		
V2		9			9			9			9		3	1	5		6	
V3													1/3	1/5	1			
VI1										1/2			1/5			1	1/5	3
VI2											4			1/6		5	1	7
VI3																1/3	1/7	1

For pairwise comparison values in an SPCM, an error model is assumed as follows:

$$r_{c'a'}^{ca} = \varepsilon_{c'a'}^{ca} \frac{v_{ca}}{v_{c'a'}} \tag{3}$$

Taking the logarithms of both sides gives

$$\log r_{c'a'}^{ca} = \log v_{ca} - \log v_{c'a'} + \log \varepsilon_{c'a'}^{ca} \tag{4}$$

To simplify the equation, logarithms will be represented by overdots as $\dot{r}_{c'a'}^{ca} = \log r_{c'a'}^{ca}, \dot{v}_{ca} = \log v_{ca}, \dot{\varepsilon}_{c'a'}^{ca} = \log \varepsilon_{c'a'}^{ca}$. Using this notation, Eq. (4) becomes

$$\dot{r}_{c'a'}^{ca} = \dot{v}_{ca} - \dot{v}_{c'a'} + \dot{\varepsilon}_{c'a'}^{ca}, c, c' = 1, \ldots, C, a, a' = 1, \ldots, A \tag{5}$$

From Eqs. (1) and (2), we have

$$\dot{r}_{c'a'}^{ca} = -\dot{r}_{ca}^{c'a'} \tag{6}$$

$$\dot{r}_{ca}^{ca} = 0 \tag{7}$$

If $\varepsilon_{c'a'}^{ca}$ is assumed to follow an independent probability distribution of mean 0 and variance σ^2, irrespective of c, a, c', a', the least squares estimate gives the best estimate for the error model of Eq. (5) according to the Gauss Markov theorem.

Equation (5) comes to following Eq. (8) by vector notation.

$$\dot{\mathbf{Y}} = \mathbf{S}\dot{\mathbf{x}} + \dot{\varepsilon} \qquad (8)$$

Where

$$\dot{\mathbf{x}} = (\ \dot{v}_{I2}\ \ \dot{v}_{I3}\ \ \dot{v}_{II1}\ \ \dot{v}_{II2}\ \ \dot{v}_{II3}\ \ \dot{v}_{III1}\ \ \dot{v}_{III2}\ \dot{v}_{III3}\ \dot{v}_{IV1}\ \bullet\ \ \bullet\ \ \bullet\ \ \dot{v}_{VI2}\ \dot{v}_{VI3})^{\mathrm{T}},$$

$$
\dot{\mathbf{Y}} = \begin{pmatrix}
\dot{r}_{I2}^{II} \\
\dot{r}_{I3}^{II} \\
\dot{r}_{III1}^{II} \\
\dot{r}_{IIII}^{I} \\
\dot{r}_{IV1}^{I} \\
\dot{r}_{I3}^{I2} \\
\dot{r}_{II2}^{III} \\
\dot{r}_{II3}^{III} \\
\dot{r}_{IIII}^{III} \\
\dot{r}_{IIV1}^{III} \\
\dot{r}_{II3}^{II2} \\
\dot{r}_{III2}^{IIII} \\
\dot{r}_{III3}^{IIII} \\
\dot{r}_{IV1}^{IIII} \\
\dot{r}_{III3}^{III2} \\
\bullet \\
\bullet \\
\bullet \\
\dot{r}_{VI3}^{VI2}
\end{pmatrix}
=
\begin{pmatrix}
\log(1/3) \\
\log 5 \\
\log(1/3) \\
\log 3 \\
\log(1/3) \\
\log 3 \\
\log 7 \\
\log 3 \\
\log 3 \\
\log 1 \\
\log(1/3) \\
\log 1 \\
\log(1/3) \\
\log(1/3) \\
\log(1/3) \\
\bullet \\
\bullet \\
\bullet \\
\log 7
\end{pmatrix}
, \ \mathbf{S} =
\begin{pmatrix}
-1 & & & & & & & & \\
& -1 & & & & & & & \\
& & -1 & & & & & & \\
& & & -1 & & & & & \\
& & & & -1 & & & & \\
1 & -1 & & & & & & & \\
1 & & -1 & & & & \bullet\ \bullet\ \bullet & \\
1 & & & -1 & & & & \\
1 & & & & -1 & & \\
1 & & & & & -1 & \\
& 1 & -1 & & & & \\
& & 1 & -1 & & & \\
& & 1 & & -1 & & \\
& & 1 & & & -1 & \\
& & & 1 & -1 & & \\
& & & & & \bullet & & \bullet \\
& & & & & \bullet & \bullet \\
& & & & & \bullet & & & \bullet \\
& & & & & & & & 1 & -1
\end{pmatrix}
$$

To simplify calculations, $v_{11} = 1$, that is $\dot{v}_{11} = 0$. The least squares estimates for formula (8) are calculated by $\hat{\mathbf{X}} = (\mathbf{S}^T\mathbf{S})^{-1}\mathbf{S}^T\dot{\mathbf{Y}}$.

4 Conclusion

SPCM of MDAHP is an incomplete pairwise comparison matrix. Therefore, the LLSM based on an error model or an eigenvalue method such as the Harker method or two-stage method is applicable to the calculation of evaluation values from an SPCM. In this paper, we showed the way of using SPCM with the application of LLSM for calculation of MDAHP with hierarchical criteria.

Table 1 shows the evaluation values obtained from the SPCM for this example.

Table 1. Evaluation values obtained by SPCM+LLSM

Criterion	I	II	III	IV	V	VI	Overall evaluation value
Alternative 1	1	2.859	0.491	2.536	4.765	1.071	12.723
Alternative 2	1.616	0.492	1.113	0.702	9.403	2.152	15.479
Alternative 3	0.328	1.186	2.219	0.597	1.728	0.506	6.564

References

1. Saaty, T.L.: The Analytic Hierarchy Process. McGraw-Hill, New York (1980)
2. Kinoshita, E., Nakanishi, M.: Proposal of new AHP model in light of dominative relationship among alternatives. J. Oper. Res. Soc. Japan **42**, 180–198 (1999)
3. Kinoshita, E., Sekitani, K., Shi, J.: Mathematical Properties of Dominant AHP and Concurrent Convergence Method. J. Oper. Res. Soc. Japan **45**, 198–213 (2002)
4. Ohya, T., Kinoshita, E.: Proposal of Super Pairwise Comparison Matrix. In: Watada, J., et al. (eds.) Intelligent Decision Technologies, pp. 247–254. Springer, Heidelberg (2011)
5. Ohya, T., Kinoshita, E.: Super Pairwise Comparison Matrix in the Multiple Dominant AHP. In: Watada, J., et al. (eds.) Intelligent Decision Technologies. Smart Innovation, Systems and Technologies 15, vol. 1, pp. 319–327. Springer, Heidelberg (2012)
6. Ohya, T., Kinoshita, E.: Super pairwise comparison matrix with the logarithmic least squares method. In: Neves-Silva, R., et al. (eds.) Intelligent Decision Technologies. Frontiers in Artificial Intelligence and Applications, vol. 255, pp. 390–398. IOS press (2013)
7. Ohya, T., Kinoshita, E.: The Treatment of Hierarchical Criteria in Dominant AHP with Super Pairwise Comparison Matrix. In: Neves-Silva, R., et al. (Eds.) Smart Digital Futures 2014, pp. 142–148. IOS press (2014)
8. Ohya, T., Kinoshita, E.: Using Super Pairwise Comparison Matrix for Calculation of the multiple dominant AHP. In: Neves-Silva, R., et al. (eds.) Intelligent Decision Technologies, Smart Innovation, Systems and Technologies 39, pp. 493–499. Springer, Heidelberg (2015)
9. Ohya, T., Kinoshita, E.: Super Pairwise Comparison Matrix in the Dominant AHP. In: Czarnowski, I., et al. (eds.) Intelligent Decision Technologies 2016, Smart Innovation, Systems and Technologies 57, pp. 407–414. Springer, Heidelberg (2016)
10. Harker, P.T.: Incomplete pairwise comparisons in the Analytic Hierarchy Process. Math. Model. **9**, 837–848 (1987)

A Questionnaire Method of Class Evaluations Using AHP with a Ternary Graph

Natsumi Oyamaguchi[1]([⊠]) [iD], Hiroyuki Tajima[1] [iD],
and Isamu Okada[2] [iD]

[1] Shumei University, Yachiyo City, Chiba, Japan
{p-oyamaguchi, tajima}@mailg.shumei-u.ac.jp
[2] Soka University, Hachioji City, Tokyo, Japan
okada@soka.ac.jp

Abstract. One of the primary activities for the faculty development of universities is to improve contents and methodology of classes by questionnaire surveys for students. In this study, we propose AHP with a ternary graph to a questionnaire survey for class evaluation in order to measure students' integrating evaluations which reflect on their concepts of values. Our method has two features. First, a weight vector for three criteria: Expertise, clarity, and personality, is measured not by calculating three combinations of pairwise comparisons but by selecting a grid area in a ternary graph. The proposed method is favorable for anonymous questionnaire surveys because respondents cannot be identified, and thus it is impossible to ask the respondent to reassess the pairwise comparisons if the C.I. value exceeds a threshold. Second, the weighting of the criteria on the goal and that of the alternatives on the criteria are by different two respondents. The former is by the students while the latter is by a faculty member. This division reflects on not only the easiness of assessments but also the consistency with the objective of the class evaluation. In the future work, we can develop a matching system which recommends the suitable classes for students in combination with needs of the students and characteristics of the classes. This is because a student may favor a class favored by the previous students who have similar concept of views on class evaluation.

Keywords: Class evaluation · Questionnaire survey
Analytic Hierarchy Process · Ternary graph

1 Introduction

The faculty development of universities in Japan has been regarded as extremely important from the point of human resource development. One of the primary activities of the faculty development is to improve contents and methodology of classes by questionnaire surveys for students, and thus many universities not only activate their improvements but also many academic papers support effective faculty development. The main purpose of questionnaire surveys is to assess faculty members. The results analyzing the surveys should be considered for the improvement of classes and learning outcomes for the students [1, 2].

© Springer International Publishing AG, part of Springer Nature 2019
I. Czarnowski et al. (Eds.): KES-IDT 2018, SIST 97, pp. 173–180, 2019.
https://doi.org/10.1007/978-3-319-92028-3_18

To contribute an improvement, questionnaire surveys must measure needs of students promptly and must communicate their results with faculty members with certainty. However, our teaching experiences suggest that there often exist mismatches of integrating evaluations and specified questions. For example, a student gave a low degree for an overall judgment while she gave high degrees for many specified points. In contrast to this, another student gave a high degree for the overall judgment while he gave low degrees for many specified points. The simple average of specified points cannot give a sufficient summary of the survey, and thus we need another approach to explain such a mismatch.

In a general approach on questionnaire survey of class evaluation, the overall evaluation is calculated as an arithmetic mean of evaluations for all specified questions. This calculation cannot reflect what evaluation an evaluator weighs heavily. Even if two evaluators give the same degree for a same question, the degree may have different weights for them.

An Analytic Hierarchy Process (AHP) [3, 4] is favorable to such a situation. AHP can quantify subjective elements including concepts of values. By installing AHP into the questionnaire survey, students' integrating evaluations which reflect on their concepts of values can be measured, and thus a mismatch between specified questions and the overall question can be resolved [5, 6].

Here we propose a method using a ternary graph [7, 8]. This is effective to measure a weight vector of three elements because students consider only what they select an area in a ternary graph. In a general usage of AHP, three pairwise comparisons are needed for three elements because $_3C_2 = 3$. Moreover, the C.I. value of the comparison matrix should be less than a threshold. If not, additional pairwise comparisons are required to an evaluator. However, it may be impossible if the questionnaire survey should be done anonymously.

2 Methods

In this study, we use a hierarchical diagram of AHP as shown in Fig. 1. In this AHP, the goal is to evaluate a class and the criteria are three elements: Expertise, clarity, and personality. These criteria were extracted by our exhaustive surveys of 19 universities in Japan. Their meanings are, respectively, whether students gained knowledge of the Expertise, whether they understood the content presented by their teacher, and whether their teacher had a personal attractiveness. The alternatives of the AHP are the following seven elements which are also extracted by our exhaustive surveys.

1. The theme of the class was always shown clearly.
2. Contents wrote on the blackboard and supplementary materials were easy for you to understand.
3. The teacher explained the contents clearly as you understood.
4. The teacher taught with eagerness.
5. The teacher taught in a suitable difficulty for your understanding.
6. The teacher promptly considered questions and comments in and out of class.
7. The teacher had a sufficient expertise.

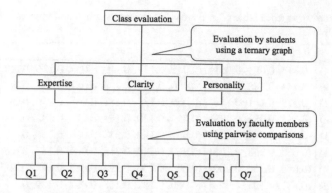

Fig. 1. A hierarchical diagram

In the ternary graph, when an evaluator indicates an area close to a vertex, V, she weighs V relatively heavily. In the pioneering work by [7], an evaluator should mark a dot in the triangle. However, this method has two constraints, i.e., she never marks on the border, and it is difficult to accurately measure the point she marks, in exchange of no constraint for plotting. In contrast to their method, we propose another method using a ternary graph as shown in Fig. 2.

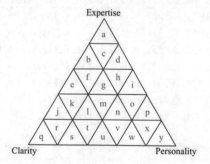

Fig. 2. A ternary graph for the questionnaire survey.

As shown in the figure, the triangle is divided into 25 areas using a ternary graph. A vector $c_R = (Re, Rc, Rp)$ is uniquely determined for a grid area $R(R \in \{a,b,c,...y\})$, where Re, Rc, and Rp denotes, the value of Expertise, that of clarity, and that of personality, respectively, where those values are between 1 to 5. A correspondence table of grid areas and their weight vectors, v_R, calculated by a pairwise comparison matrices are shown as Table 1, and the meanings of values in a pairwise comparison matrix are shown in Table 2. For an example, the case of the grid area "h" is shown in Table 3.

Table 1. A correspondence table of grid areas in the ternary graph and their weight vectors. w_R is defined later.

R	C_R	V_R	W_R
a	(5, 1, 1)	(0.818, 0.091, 0.091)	(0.059, 0.133, 0.179, 0.057, 0.126, 0.086, 0.361)
b	(4, 2, 1)	(0.731, 0.188, 0.081)	(0.060, 0.141, 0.199, 0.057, 0.131, 0.086, 0.327)
c	(4, 1, 1)	(0.778, 0.111, 0.111)	(0.059, 0.134, 0.181, 0.061, 0.129, 0.090, 0.346)
d	(4, 1, 2)	(0.731, 0.081, 0.188)	(0.059, 0.130, 0.169, 0.077, 0.132, 0.104, 0.328)
e	(3, 3, 1)	(0.455, 0.455, 0.091)	(0.061, 0.162, 0.253, 0.066, 0.148, 0.093, 0.218)
f	(3, 2, 1)	(0.637, 0.258, 0.105)	(0.060, 0.146, 0.211, 0.064, 0.137, 0.092, 0.290)
g	(3, 2, 2)	(0.600, 0.200, 0.200)	(0.060, 0.140, 0.192, 0.082, 0.141, 0.109, 0.276)
h	(3, 1, 2)	(0.637, 0.105, 0.258)	(0.060, 0.131, 0.168, 0.092, 0.139, 0.1 19, 0.291)
i	(3, 1, 3)	(0.455, 0.091, 0.455)	(0.061, 0.127, 0.149, 0.133, 0.152, 0.157, 0.222)
J	(2, 4, 1)	(0.188, 0.731, 0.081)	(0.063, 0.185, 0.309, 0.071, 0.164, 0.096, 0.1 12)
k	(2, 3, 1)	(0.258, 0.637, 0.105)	(0.062, 0.177, 0.288, 0.074, 0.160, 0.099, 0.140)
l	(2, 3, 2)	(0.200, 0.600, 0.200)	(0.063, 0.172, 0.273, 0.093, 0.165, 0.1 17, 0.118)
m	(2, 2, 2)	(0.333, 0.333, 0.333)	(0.062, 0.148, 0.207, 0.1 14, 0.158, 0.138, 0.172)
n	(2, 2, 3)	(0.200, 0.200, 0.600)	(0.062, 0.133, 0.159, 0.167, 0.169, 0.188, 0.123)
o	(2, 1, 3)	(0.258, 0.105, 0.637)	(0.062, 0.125, 0.136, 0.172, 0.166, 0.193, 0.146)
P	(2, 1, 4)	(0.188, 0.081, 0.731)	(0.062, 0.121, 0.124, 0.191, 0.172, 0.21 1, 0.119)
q	(1, 5, 1)	(0.091, 0.818, 0.091)	(0.063, 0.192, 0.326, 0.075, 0.170, 0.100, 0.074)
r	(1, 4, 1)	(0.111, 0.778, 0.111)	(0.063, 0.188, 0.316, 0.079, 0.169, 0.103, 0.082)
s	(1, 4, 2)	(0.081, 0.731, 0.188)	(0.063, 0.183, 0.300, 0.094, 0.172, 0.1 17, 0.071)
t	(1, 3, 2)	(0.105, 0.637, 0.258)	(0.063, 0.174, 0.275, 0.106, 0.171, 0.129, 0.081)
u	(1, 3, 3)	(0.091, 0.455, 0.455)	(0.063, 0.156, 0.222, 0.143, 0.175, 0.164, 0.078)
v	(1, 2, 3)	(0.105, 0.258, 0.637)	(0.063, 0.137, 0.167, 0.176, 0.176, 0.196, 0.085)
w	(1, 2, 4)	(0.081, 0.188, 0.731)	(0.063, 0.130, 0.145, 0.194, 0.178, 0.213, 0.077)
x	(1, 1, 4)	(0.111, 0.111, 0.778)	(0.063, 0.123, 0.126, 0.202, 0.177, 0.221, 0.089)
y	(1, 1, 5)	(0.091, 0.091, 0.818)	(0.063, 0.120, 0.118, 0.210, 0.178, 0.228, 0.082)

Table 2. The meanings of values using in a pairwise comparison matrix.

A differences of distances from two vertices	Definition	Value
0	Equal importance	1
1	Weak importance	3
2	Strong importance	5
3	Very strong importance	7
4	Absolute importance	9

Table 3. A pairwise comparison matrix of the grid area "h".

	Expertise	Clarity	Personality
Expertise	1	5	3
Clarity	1/5	1	1/3
Personality	1/3	3	1

The pairwise comparisons for weighting three criteria are conducted by a faculty member. The weight vectors of pairwise comparison matrices of alternatives are v_e, v_c and v_p. With vectors v_e, v_c and v_p, a matrix S is defined as

$$S = [v_e, v_c, v_p].$$

Using those three vectors, the final weight of the hierarchy is yielded as

$$w_R = S \cdot {}^t v_R.$$

Here ${}^t v_R$ means the transposed matrix of v_R. Summing up these processes, the overall values of the hierarchy in Fig. 1 can be calculated. This calculation follows the general method of AHP.

The vector $s_n = (Q1_n, Q2_n, Q3_n, Q4_n, Q5_n, Q6_n, Q7_n)$ represents a sequence of values of the seven questions answered by each student n ($n \in \{1, 2, \ldots 25\}$). Using above weights, we can compute students' integrating evaluations which reflect their concepts of values as v_n. The overall value v_n of student n is defined with s_n ($n \in \{1, 2, \ldots 25\}$) and w_R (R denotes the area student n chose) as follows:

$$v_n = s_n \cdot w_R.$$

3 Preliminary Results

In this paper, we conducted a questionnaire survey of 27 freshman students of the Department of Mathematics, School of Teacher Education, Shumei University (age: 19 or 20, 9 female and 18 male). The number of valid responses was 25. The aggregate result of selecting grid areas of the ternary graph is shown in Fig. 3. In the diagram, students answered the following question: We have three evaluating factors: Expertise, clarity, and personality. Please select the most suitable grid area out of the 25 considering weights you give to these three factors. Table 4 and Fig. 3 show the students' responses.

The pairwise comparisons for weighting three criteria were conducted by the first author as shown in Table 5.

The eigenvalues and the weight vectors of pairwise comparison matrices of alternatives are calculated as $\lambda_{max,e} = 7.582$, $v_e = {}^t(0.058, 0.127, 0.168, 0.035, 0.114, 0.066, 0.432)$, $\lambda_{max,c} = 7.551$, $v_c = {}^t(0.064, 0.208, 0.370, 0.061, 0.175, 0.086, 0.037)$, and $\lambda_{max,p} = 7.465$, $v_p = {}^t(0.063, 0.110, 0.085, 0.246, 0.186, 0.262, 0.048)$, respectively. Each C.I. value of the matrix is, in order, 0.097, 0.092, and 0.078. Neither exceeds the threshold (0.1). Resulting weights of each grid area w_R are shown in the rightmost column in Table 1. Finally, students' integrating evaluations v_n which reflect on their concepts of values are shown in the rightmost column in Table 4.

Table 4. The questionnaire results of 25 students. Q1 to Q7 corresponds to the alternatives of the AHP previously explained. The values of alternatives are, respectively, 5 (strongly think so), 4 (think so), 3 (do no know or do not care), 2 (do not think so), and 1 (never think so).

Student number	Grid area selected	Q1	Q2	Q3	Q4	Q5	Q6	Q7	v_n
1	m	5	5	5	5	5	5	5	5.00
2	m	4	4	4	4	4	4	4	4.00
3	n	4	4	5	5	5	5	5	4.81
4	m	5	5	5	5	4	5	5	4.84
5	r	5	5	5	5	5	5	5	5.00
6	g	5	5	4	5	5	5	4	4.53
7	m	5	5	5	5	5	5	4	4.82
8	n	5	4	5	5	4	4	4	4.39
9	n	5	5	5	5	4	5	5	4.83
10	r	4	4	4	5	4	4	5	4.16
11	n	4	4	5	5	5	5	4	4.68
12	l	5	5	5	5	3	5	5	4.67
13	u	4	4	3	4	2	4	4	3.43
14	m	5	5	5	4	3	5	5	4.57
15	m	5	4	5	5	5	5	3	4.50
16	h	4	5	5	5	5	5	5	4.94
17	m	5	5	4	5	4	5	4	4.46
18	k	3	4	4	3	4	4	3	3.73
19	m	4	5	5	4	5	5	5	4.82
20	l	4	4	3	4	4	4	3	3.61
21	g	4	5	3	4	3	3	4	3.70
22	f	4	5	4	4	2	4	4	3.87
23	m	4	5	4	4	4	4	4	4.15
24	t	4	5	4	5	3	5	3	4.16
25	m	4	4	4	5	4	4	4	4.11

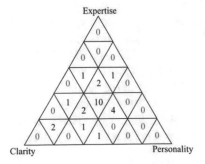

Fig. 3. The number students select each grid area in the ternary graph.

Table 5. A pairwise comparison matrix of alternatives on (a) Expertise, (b) clarity, and (c) personality.

(a)	Q1	Q2	Q3	Q4	Q5	Q6	Q7	(b)	Q1	Q2	Q3	Q4	Q5	Q6	Q7	(c)	Q1	Q2	Q3	Q4	Q5	Q6	Q7
Q1	1	1/2	1/3	2	1/3	1	1/5	Q1	1	1/3	1/9	3	1/3	1/3	1	Q1	1	1	1	1/5	1/3	1/5	1
Q2	2	1	1	5	2	1	1/5	Q2	3	1	1	3	1	3	5	Q2	1	1	1	1/3	1	1/5	5
Q3	3	1	1	5	3	3	1/5	Q3	9	1	1	7	3	5	7	Q3	1	1	1	1/3	1/3	1/3	3
Q4	1/2	1/5	1/5	1	1/5	1/2	1/5	Q4	1/3	1/3	1/7	1	1/3	1	3	Q4	5	3	3	1	1	1	5
Q5	3	1/2	1/3	5	1	3	1/5	Q5	3	1	1/3	3	1	3	5	Q5	3	1	3	1	1	1	3
Q6	1	1	1/3	2	1/3	1	1/5	Q6	3	1/3	1/5	1	1/3	1	3	Q6	5	5	3	1	1	1	3
Q7	5	5	5	5	5	5	1	Q7	1	1/5	1/7	1/3	1/5	1/3	1	Q7	1	1/5	1/3	1/5	1/3	1/3	1

4 Discussion

In this study, we propose an AHP method using a ternary graph for a questionnaire survey for class evaluation in order to measure the needs of the students correctly. Our method has two features. First, a weight vector for three elements is measured by not three combinations of pairwise comparisons but selecting a grid area in a ternary graph. The anonymous questionnaire cannot identify a respondent, and thus it is impossible to reassess the comparisons if the C.I. value exceeds a threshold. To avoid this point, selecting a grid area in the ternary graph can be uniquely identified the closest weight vector whose C.I. value doesn't exceed a threshold. The merit of this method is to measure the assessments responded by students to consider their individual weights of each element. On the other hand, selecting a grid area in the ternary graph is a difficult task for the students, and thus this point requires a future consideration.

Second, the weighting of the criteria on the goal and that of the alternatives on the criteria are by two different respondents. The former is by the students while the latter is by a faculty member. The pairwise comparisons among seven items require 21 comparisons. Not only this point, but also the former weighting is suitable not for the students but faculty members because their items are designed by the faculty members themselves.

In future work, this analysis will be useful for the future students to choose their classes by aggregating data. This method expects that the needs of the students and the characteristics of the classes can be measured, and so we can develop a matching system which recommends the suitable classes for students. The matching system considers the information of a grid area (one of 25) a student selects in the ternary graph, because such information is regarded as a reflection on one's concept of views. Then, the previous students who had the same concepts of views are extracted. It is not so unnatural that the student and the previous ones have similar concept of views on class evaluation, and thus the student may favor a class favored by the previous students.

References

1. Cohen, E.H.: Student evaluations of course and teacher: factor analysis and SSA approaches. Assess. Eval. High. Educ. **30**(2), 123–136 (2005)
2. Toland, M.D., De Ayala, R.J.: A multilevel factor analysis of students' evaluations of teaching. Educ. Psychol. Measur. **65**(2), 272–296 (2005)
3. Saaty, T.L.: A scaling method for priorities in hierarchical structures. J. Math. Psychol. **15**(3), 234–281 (1977)
4. Saaty, T.L.: The Analytic Hierarchy Process. McGraw-Hill, New York (1980)
5. Ikehara, K., Toyoda, H.: A course evaluation model using paired comparisons to integrate importance of rating criteria and students' ratings of teaching: comparison of students' and teachers' evaluations. Jpn. J. Educ. Psychol. **60**(1), 48–59 (2012)
6. Zhao, C., Zhao, Y., Tan, X., Li, Y., Luo, L., Xiong, Y.: Course evaluation method based on analytic hierarchy process. Future Commun. Comput. Control Manag. **2**, 275–283 (2012)
7. Mizuno, T., Taji, K.: A ternary diagram and pairwise comparisons in AHP. J. Jpn. Symp. Anal. Hierarchy Process **4**, 59–68 (2015)
8. Mizuno, T., Taji, K.: Analyses of pairwise comparisons with a ternary diagram. In: ISAHP 2016, Creative Decision Foundation on behalf of the International Symposium on the Analytic Hierarchy Process, Pittsburgh (2016)

A Link Diagram for Pairwise Comparisons

Takafumi Mizuno[⊠]

Meijo University, Minami-Yada 4-102-9, Higashi-ku, Nagoya, Japan
tmizuno@meijo-u.ac.jp

Abstract. In decision making processes such as Analytic Hierarchy Process, weights of alternatives are derived from pairwise comparisons. I propose a link diagram to visualize pairwise comparisons and weights. In the diagram, each alternative is represented in a loop floating on the ground. Weights of alternatives are represented as heights of corresponding loops from the ground, and crossing of loops represents which alternative is superior or inferior. So, decision-makers can grasp roughly inconsistency of pairwise comparisons just by viewing the diagram.

Keywords: Pairwise comparisons · Visualization · Link diagram

1 Pairwise Comparisons and Deriving Weights

Pairwise comparisons are important procedures for decision-makings. Decision-makers often evaluate alternatives on intangible criteria quantitatively by comparing them directly. Analytic Hierarchy Process (AHP) is a process to solve multi-criteria decision-making problem with doing pairwise comparisons. AHP has been proposed by Saaty [4] in late 1970s. The process consists of three phases: construction of the hierarchical structure, deriving weights of alternatives from pairwise comparisons, and synthesis of the weights.

The hierarchical structure, which is typically three layered, represents the decision-making problem. The top layer represents the purpose of the problem which a decision-maker want to solve, the middle layer represents the set of criteria $\{c_1, \cdots, c_m\}$, and the bottom layer represent the set of alternatives $\{X_1, \cdots, X_n\}$.

By pairwise comparisons, the decision-maker shows ratios of elements and derive weights from the ratios. Let m and n be the numbers of criteria and alternatives, respectively. Let $b_j \in \mathbb{R}_{>0}$ be the weight of j-th criterion to achieve the purpose, and let $E_{ij} \in \mathbb{R}_{>0}$ be the weight of i-th alternative in regard with j-th criterion.

In the synthesis phase, the score of i-th alternative is calculated as follows.

$$S_i = \sum_{j=1}^{m} E_{ij} b_j. \tag{1}$$

© Springer International Publishing AG, part of Springer Nature 2019
I. Czarnowski et al. (Eds.): KES-IDT 2018, SIST 97, pp. 181–186, 2019.
https://doi.org/10.1007/978-3-319-92028-3_19

The decision-maker chooses the alternative with the highest score.

Pairwise comparison is the most critical procedure in AHP. The decision-maker picks a pair (X_i, X_j) up from the set of alternatives $\{X_1, \cdots, X_n\}$, and judges which alternative is more important. The judgement is represented in the ratio a_{ij} that means X_i is a_{ij} times more important as X_j. The decision-maker gives ratios for all pairs, and arranges the ratios into the matrix A as follows.

$$A = \begin{bmatrix} 1 & a_{12} & \cdots & a_{1n} \\ a_{21} & 1 & \cdots & a_{2n} \\ \vdots & & \cdots & \vdots \\ a_{n1} & a_{n2} & \cdots & 1 \end{bmatrix}. \tag{2}$$

The matrix is referred to as *pairwise comparison matrix*. Each element a_{ij} has reciprocal property; $a_{ij} = 1/a_{ji}$ for $i, j \in \{1, ..., n\}$.

There are some procedures to derive weights of alternatives from the pairwise comparison matrix. In AHP, the eigenvector method and the geometric mean method are often used. Let $\lambda_{\max}$ and $\boldsymbol{w} = [w_1, \ldots, w_n]^\top$ be the principal eigenvalue and the principal eigenvector of the matrix A in (2). They satisfy

$$A\boldsymbol{w} = \lambda_{\max}\boldsymbol{w} \text{ and } \sum_{i=1}^{n} w_i = 1. \tag{3}$$

The eigenvector method adopts i-th element w_i of $\boldsymbol{w}$ as the weight of X_i. The geometric mean method calculates the geometric mean of i-th row of the matrix as the weight of X_i.

It is considered that these ratios may contain perturbations or errors of judgements of the decision maker. Consistency Index (C.I.) is widely used to measure how tolerable to use results of pairwise comparisons. C.I. value of the matrix A in (2) is calculated as

$$CI = \frac{\lambda_{\max} - n}{n - 1}. \tag{4}$$

In practical use of AHP, it is considered that when C.I. value is less than 0.1 or 0.15, the results of the pairwise comparisons are tolerable; otherwise, redoing pairwise comparisons is recommended.

C.I. value is zero if and only if the matrix A is completely consistent that means

$$a_{ij}a_{jk} = a_{ik}, \quad i, j, k \in \{1, \ldots, n\}. \tag{5}$$

When the pairwise comparison matrix is completely consistent, the rank of the matrix is 1, and each column represents the weights of alternatives [5]. Then all weights satisfy $w_i/w_j = a_{ij}$, $i, j \in \{1, ..., n\}$.

C.I. value is easy to calculate and evaluates consistency of whole comparisons, but it cannot indicate which comparison is irrational. Directed graphs are often used to visualize pairwise comparisons or to reduce inconsistency of them [2,3]. Each node of graphs represents an alternative, and each edge connects from an alternative to more important alternative. Directed graphs, however, are not suitable to visualize pairwise comparisons. Increasing alternatives increases edges, and decision makers cannot grasp relations between alternatives easily. And it is hard to see on the graphs simultaneously pairwise comparisons and weights derived from them.

In this article, I propose a link diagram which represents pairwise comparisons and weights. Using the diagram makes decision-makers see all comparisons easily, and grasp relations between derived weights and the comparisons. Moreover, in the diagram, decision makers can see inconsistency of comparisons roughly.

2   A Link Diagram

A main idea of this article is that represents each alternative as a loop floating on the ground. If seeing weights of alternatives is need, the weights are represented as heights of the loops. Superiority or inferiority of a pair of alternatives is represented as how loops crossing. In the crossing, the loop of superior alternative is above the loop of inferior one. At whole the diagram, each loop crosses another loops twice. One crossing represents a result of pairwise comparison, and other represents order of derived weight.

There is an example for three alternatives. Results of pairwise comparisons are represented in the matrix as follows.

$$A = \begin{bmatrix} 1 & 1/2 & 3 \\ 2 & 1 & 1/2 \\ 1/3 & 2 & 1 \end{bmatrix}. \tag{6}$$

In the example, there is a cycle of preferences; $X_3 \succ X_2 \succ X_1 \succ X_3$.

The link diagram consists of two blocks: a block representing pairwise comparisons, and another block representing derived weights (Fig. 1). First, I describe how to represent pairwise comparisons at the first block. Since pairwise comparison matrices are reciprocal symmetric, I consider only elements above the main diagonal of the pairwise comparison matrix. At the crossing of X_1 and X_2, X_2 is above X_1, because $(1,2)$-elements is $1/2$. Similarly, X_1 is above X_3 at the crossing of X_1 and X_3, and X_3 is above X_2 at the crossing X_2 and X_3.

Next, I describe how to represent weights derived from the matrix at the second block. Weights, derived by the principal eigenvector method, satisfy $w_1 : w_2 : w_3 = \sqrt[3]{3/2} : 1 : \sqrt[3]{2/3}$, and it means that $X_1 \succ X_2 \succ X_3$. In the block, X_1 is above all loops at all crossings. X_2 is beneath X_1 at crossing of X_1 and X_2, and is above X_3 at crossing X_2 and X_3. X_3 is beneath at all loops at all crossings.

A contradiction, if it occurs, can be grasp easily by seeing the diagram. When a cycle of preferences occurs, there is a pair of loops cannot be splittable each other even if how decide weights.

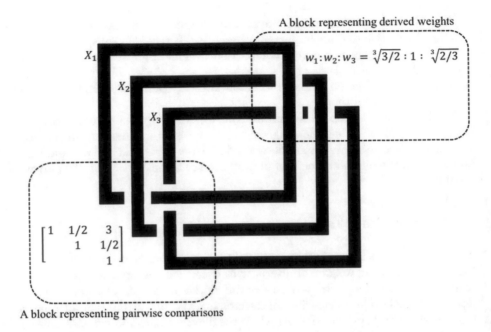

Fig. 1. A link diagram which represents the pairwise comparison matrix in (6) and weights derived from the matrix

To grasp inconsistency of pairwise comparisons, I represent weights as heights of loops from the ground. In the block of derived weights, each height of loop is fixed to be its weight. While, in the block of pairwise comparisons, ratios of weights are represented as ratios of heights of loops. The height of loop X_3 is fixed whole the block; the height is w_3. For the loop X_2, at the point crossing to X_3 (p_{23} in Fig. 2), the ratio of height of X_2 and height of X_3 is $a_{23} = 1/2$. Or the height of X_2 at the point determined to be $a_{23}w_3$. Except the point, height of X_2 is fixed to be w_2. For the loop X_1, at the point crossing to X_2 (p_{12}), the ratio of the height of X_1 and the heights X_2 is $a_{12} = 1/2$, or the height at the point is $a_{12}w_2$. At the point crossing to X_3 (p_{13}), the ratio of the height of X_1 and the height of X_3 is $a_{13} = 3$, or the height at the point is $a_{13}w_3$. Except the two points (p_{12} and p_{13}), heights of X_1 is fixed to be w_1. In cases of more alternatives, every heights of crossing points can be determined similarly backward from X_n to X_1. Consistency of pairwise comparisons are grasp roughly by viewing the diagram from view lines parallel to the ground (Figs. 2 and 3).

If pairwise comparisons are completely consistent, then the diagram is seen as line segments parallel to the ground.

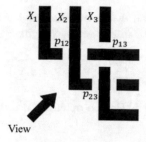

Fig. 2. A view point to check consistency of the pairwise comparisons of Fig. 1.

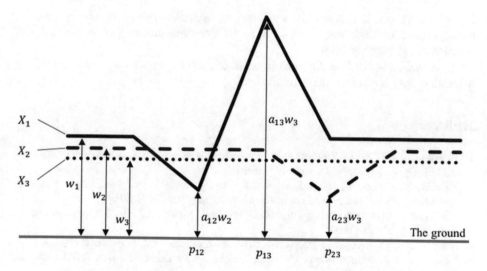

Fig. 3. The link diagram seen from view point indicated in Fig. 2.

3   Conclusions

I proposed a link diagram to represent results of pairwise comparisons. The diagram makes decision-makers see simultaneously pairwise comparisons and weights.

From mathematical view, the link diagram consists of components whose number equals to the number of alternatives. I mentioned that the link diagram has a pair of loops not splittable when cycles of preference occur. It seems to be trivial, but I did not give its proof in this article.

I represented alternatives as loops floating on the ground to grasp inconsistency of pairwise comparisons. It visualizes relation between w_i and $a_{ij}w_j$, $j \neq i$.

Deriving weights from pairwise comparison matrices is considered as an optimization problems [1,2,6]: to find a set of weights $\{w_1, \cdots, w_n\}$, where

$$\min_{w_1, \cdots, w_n \in \mathbb{R}_+} \left| \frac{w_i}{w_j} - a_{ij} \right|, \quad i, j \in \{1, \cdots, n\}. \tag{7}$$

If the solution $\{w_1, \cdots, w_n\}$ is Pareto optimal, it holds the statement: for all set of weights $\{w_1', \cdots, w_n'\}$,

$$\forall i, j \in \{1, \cdots, n\}, \exists l, m \in \{1, \cdots, n\},$$
$$\left| w_i' - a_{ij} w_j' \right| < \left| w_i - a_{ij} w_j \right| \Rightarrow \left| w_l' - a_{lm} w_m' \right| > \left| w_l - a_{lm} w_m \right|. \tag{8}$$

By checking on the diagram whether w_i satisfies $\min_{j \neq i} a_{ij} w_j \leq w_i \leq \max_{j \neq i} a_{ij} w_j$ or does not for $i = 1$ to n, decision-makers can construct the solutions is Pareto optimal.

In actual uses, the link diagram consists of 3D polygons. Now, I am developing interfaces easy to use for decision makers.

References

1. Bozóki, S., Fülöp, J.: Efficient weight vectors from pairwise comparison matrices. Eur. J. Oper. Res. **264**(2), 419–427 (2018)
2. Bozóki, S., Fülöp, J., Rónyai, L.: On optimal completion of incomplete pairwise comparison matrices. Math. Comput. Modell. **52**, 318–333 (2010)
3. Nishizawa, K.: A consistency improving method in binary AHP. J. Oper. Res. Soc. Jpn. **38**(1), 21–33 (1995)
4. Saaty, T.L.: The Analytic Hierarchy Process. McGraw-Hill, New York (1980)
5. Saaty, T.L.: Fundamentals of Decision Making and Priority Theory. RWS Publication, Pittsburgh (1994)
6. Sekitani, K., Yamaki, N.: A logical interpretation for the eigenvalue method in AHP. J. Oper. Res. Soc. Jpn. **42**, 219–232 (1999)

Measurement of Abnormality in Eye Movement with Autism and Application for Detect Fatigue Level

Ippei Torii[✉], Takahito Niwa, and Naohiro Ishii

Aichi Institute of Technology, Toyota, Aichi, Japan
mac@aitech.ac.jp

Abstract. The authors aimed to establish an objective diagnostic criterion for children with autism spectrum disorder. To this end, they conducted eye-tracking tests on children with autism and neurotypical children. They obtained the pixel number variation (a numerical value) in gaze direction based on the center of mass of pixels associated with the pupil. The results were then plotted onto a two-dimensional graph, and distributions based on probability density function and receiver operating characteristic curve analysis were ascertained. This analysis yielded a decision boundary clearly demarcating autism and neurotypical distributions and, thus, confirming the reliability of the method. This finding suggests that this technique of measuring abnormality in pixel number is effective for distinguishing individuals with autism from those with typical development and, thus, can serve as an objective criterion for the diagnosis of autism. On other hand, In modern society, many people are under stress. It is over accumulated and autonomic nerves is disturbed that fatigued and it causes of mental disorders such as depression and autonomic imbalance. The evaluation of fatigue/fatigue feeling was subjectively evaluated use visual analog scale (VAS) and questionnaire paper. That means we not had way to be objectively evaluate. In this study, we will apply the system to verify fatigue level judgment by image processing and verify the relevance of fatigue degree and eye movement.

Keywords: Physically handicapped children · OpenCV
Line-of-Sight detection

1 Introduction

Autistic spectrum disorder (ASD) is a developmental disorder; the number of patients has recently been increasing rapidly due to the influence of changes in diagnostic criteria. It is also difficult to distinguish the autistic spectrum from other developmental disorders such as learning disability. Therefore, a diagnostic tool that can assess the disorder objectively and quantitatively is required.

According to a 2014 report from the United States Centers for Disease Control and Prevention, as of 2010, 1 in 68 children had been identified with ASD. This estimate is approximately 30% higher than previous estimates reported in 2012 (1 in 88 children, based on data from 2008). The sharp rise in the prevalence of ASD is related to

© Springer International Publishing AG, part of Springer Nature 2019
I. Czarnowski et al. (Eds.): KES-IDT 2018, SIST 97, pp. 187–196, 2019.
https://doi.org/10.1007/978-3-319-92028-3_20

growing public awareness of autism. When parents seek a diagnosis at an early stage, it increases the chances of detecting autism. The causes of autism are unclear and, to date, there no effective treatment or preventive measures [1, 2] (Kitazawa et al.). In Japan, approximately 10% of young boys and approximately 5% of girls have a developmental disorder. In 2003, the government proposed the Act for Support for Persons with Developmental Disabilities and, in 2006, commenced special needs education for children with disabilities [3, 4] (National Institute of Special Needs Education, Japan). In addition, the government proactively supports research activities related to developmental disabilities. Against this backdrop, there is a need for a simple, inexpensive, and quantitative diagnostic tool that anyone can use to diagnose a developmental disorder.

2 Purpose of the Study

One of the authors previously developed a technique for eye blink evaluation based on afterimages. In the present study, we used this method for measuring abnormalities in gaze direction based on variations in the center of mass of pixels associated with the pupil.

The procedure for measuring abnormality in gaze direction is as follows:

1. Identify the pixels associated with the pupil in the captured frames (detection device: OpenCV, Haar-like);
2. Prepare reference images of blink/gaze direction, known as afterimages, from the pupil pixels;
3. Place coordinates on the frame image;
4. Determine the center of mass of the pixels based on afterimages and position of the iris;
5. Subject views moving image for 10 s; during these 10 s, record coordinate changes in center of mass
6. Process camera images with a precision of 30–40 frames/s and convert the amount of change in center of mass of the pixels into numerical values.

We plotted the values yielded from the above procedure onto a graph, and then compared the typical development and autism groups using a probability density function. Based on the compared data, we calculated the decision boundary and used receiver operating characteristic (ROC) curve analysis, which confirmed the reliability of our assessment method. Thus, we developed an objective criterion that can distinguish individuals with typical (i.e., normal) development from those with autism accurately and early.

3 Structure of the System

The difference in the area of the extent of change occurs by measuring the difference between the afterimage in consecutive frames and the present frame. We set the amount of change to the quantity of eyeball movements. This method made it possible to detect and quantify changes in eyeball movements in numerical values. The difference in these movements between non-autistic and autistic individuals was compared and analyzed. Figure 1 is an illustration indicating the states to assess opening and closing of the eyes.

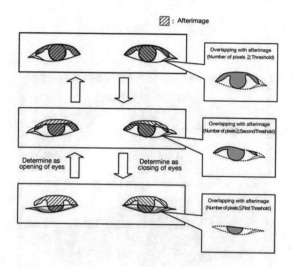

Fig. 1. Diagram of opening and closing of eye

Images are always assessed consecutively. We assigned each pixel a persistence value (p), and subtracted 1 from every frame p. Each p is an integer of zero or above. We added 5 to the p of pixels that we deemed to be black in color.

We define the black part as an afterimage based on value of image persistence. The value becomes - 1+5 when the value of image persistence is judged black, and the program is forced to consider it an afterimage when the value is >50. More than 50

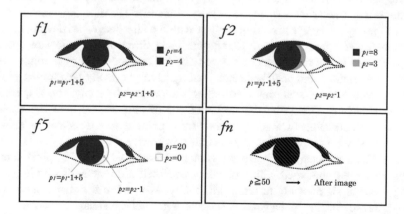

Fig. 2. The change of persistence value

Fig. 3. The center of gravity

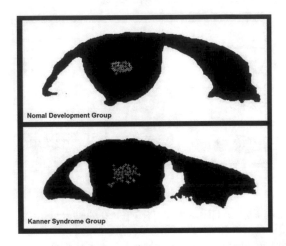

Fig. 4. Difference in eye movement detection between normal development and autism

means the part that is judged to be black during approximately 0.45 s is determined as an afterimage because 24 to 30 frames are processed per second (Fig. 2).

To detect gaze direction, there are methods using circles or ovals focused on the pupils. However, even in healthy individuals, pupils cannot be visualized unless the eyes are wide open. These techniques are not suitable for physically handicapped individuals who often cannot open their eyes sufficiently. Therefore, by considering the center of gravity of the eye and comparing it with the afterimage, we developed a new method to detect gaze direction. Using this gaze direction detection method, the average change in the pixel center of gravity of the eye movement per minute in the subject is determined and analyzed, thus distinguishing normally developing children and those with Kanner syndrome. To define the center of gravity, the acquired eye region is binarized. Generally, the first moment of circumference of the center of gravity is zero. In other words, it refers to the point where weights are balanced; therefore, this was used to calculate the center of gravity (Fig. 3).

It is difficult to accurately grasp the center of the pupil using this pixel center of gravity. However, compared with circular or oval detection methods, the image processing speed is significantly faster. Additionally, even when the shape of the pupils collapses due to blurring of the gaze, the center of gravity can always be captured. After defining the center of gravity of the pixel, movements of the pupils are detected for each frame. The average change in the quantity of this numerical value over a span of 1 min is defined as fine movement of the eyeball, and it is used to distinguish individuals in a normally developing group from those in a Autism (Kanner syndrome) (Fig. 4).

4   Clinical Application to Children with Autism Using Pixel Number Variation Abnormality

The research team from Osaka University has conducted several studies investigating eye movement abnormalities in children with attention deficit hyperactivity disorder (ADHD) using the Tobii eye tracker to measure saccadic ocular movement [5, 6] (Nakano, Kitazawa et al.). The diagnosis of such an abnormality was based on a comparison of the time the subject spent fixing their gaze on an object (Kimura) [7] (Iwanaga) [8] (Fukuda) [9], but there is no link to results. Circular movements that combine saccade and pursuit are shown in the experiment so that a child with autism could gaze. A circle that begins to move from the left moves suddenly out of the way and appears at the right. To use this as an evaluation index of autism, a method to compare and analyze data from the eyes captured by a camera in the PC to establish a density function was built. Subjects in the analysis included 37 elementary school autistic children, 15 elementary school normally developing children, 10 junior high school autistic children, 10 junior high school normally developing children, 7 high school autistic children and 5 high school normally developing children. The calculated average of the amount of change in eyeball movement with eye shift between frames was compared and analyzed. The reliability of this analysis method was measured using ROC curve analysis. Figure 5 shows the relationship between the typical-development group, the autism group, and the unable-to-follow-instructions group in terms of probability density function. The red line indicates the unable-to-follow-instructions group, the blue line indicates the typical-development group, and the green line indicates the autism group. This figure provides further confirmation of the independence of each group's data.

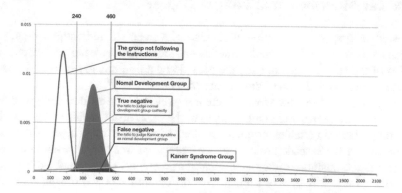

Fig. 5. Relations of normal development group and autism

From these results, the identification border between the unable-to-follow-instruction group and the normally developing group was 240. The border between the autism group and the normally developing group was 460.

Figure 6 is a truth value table showing the probabilities for positives and negatives. The probability for accurately determining typical development was 99%, and the probability of a false positive (i.e., the probability that the subject is developing normally but is actually autistic) is 1%. The probability for accurately determining autism is 96%, and the probability of a false negative (i.e., the probability that the subject is autistic but is actually neurotypical) is less than 4.0%.

	FALSE	TRUE
Negative ($\leqq 240 < 480$)	False Negative Rate 0.04	True Negative Rate 0.99
Positive ($480 \leqq$)	False Positive Rate 0.01	True Positive Rate 0.96

Fig. 6. Boolean table indication whether positive or negative

The reliability of the results is measured using ROC curve analysis [10] (Kawashima et al.). The reliability that could distinguish a normal developing individual from an autistic one is 98.77% (Fig. 15). On ROC analysis, the shape shifts to the top left where there is a small peak of two density functions, and shows that the performance of the decision surface line is high. The area under the ROC curve (AUC) was 0.9877; the closer to 1 the AUC is, the more reliable the test is. The AUC we obtained, therefore, suggests that the device produced false readings in <2% of cases. Thus, our method for detecting autism exhibits excellent accuracy.

5   Ocular Movement and Fatigue Degree

We tested to inspect the relation of ocular movement to the fatigue degree. The experiment is measure the ocular movement detection and the autonomic nerve function of the subject and compare the provided data. In this experiment 43 subjects cooperated with us and carried out the experiment.

We hypothesized that the sense of ocular movement become dull and speed to turn eyes to the object decreases in fatigue state. In the experiment of this study, we display a saccade problem in a tablet terminal and detect the ocular movement of the subject. The contents of the saccade problem that we prepared was appear a circle to become the object at the position of the random and has a subject chase an object that appear during one minute with eyes of subject. We carried out the ocular movement detection with the built-in camera of the laptop PC (Fig. 7).

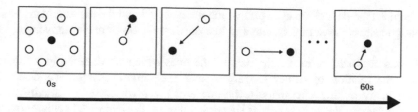

Fig. 7. The saccade program

We used the Fatigue, stress measuring system made in Hitachi systems Co., Ltd in the measurement of the autonomic nerve function. We put a forefinger in the hole of right and left of the device of the handy type and measure a pulse wave and a heart electric wave at the same time. The data which we acquired is analyzed by "the fatigue analysis server" on the cloud system of the Hitachi systems Co., Ltd. via a PC and a report is formed and an evaluation result is transmitted to a PC.

In this study, we define fatigue as a disturbed state of the autonomic nerve and we performed the fatigue judgment of the subject based on the result of a measurement of fatigue, the stress measuring system. we used LF/HF that indicate the balance of sympathetic nerve and the parasympathetic nerve from output numerical value. In addition, active mass of the autonomic nerve namely ccvTP is influenced by age, so we used the value that made ccvTP deviation value at age. The value that Hitachi systems Co., Ltd assumes the standard value that is in a normal state is LF/HF: 0.8–2.0, ccvTP deviation value: 43–57. We regard the case that did not satisfy either this standard value as fatigue and divided the subjects to the person who got tired and who did not get tired (Fig. 8). As the result, in 43, 12 people became a normal state and 31 people became fatigue state.

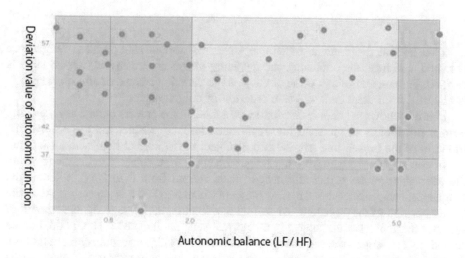

Fig. 8. Fatigue result of a measurement of the subject

We used the data of the eyes part which we detected and the measurement result of the fatigue stress measuring system and inspect the relation of ocular movement to the fatigue degree.

It was necessary to record the data of the eyes which we detected in a subject to inspect the relation of ocular movement and the fatigue degree. We detect eyes movement by putting afterimage technique and the pixel center of gravity method together when we detect eyes part as showed in before. We expressed ocular movement with numerical value by saving the position data of a pixel center of gravity provided at the time of the afterimage generation of the eyes part every 1 frame. We assumed this afterimage analysis and observed livingness of the ocular movement by the difference of the pixel center of gravity position every 1 frame. For example, we can read that an eyeball makes a gentle move when the difference of the pixel center of gravity position is small. On the contrary, it shows that eyes move quickly for an aim when the difference of the pixel center of gravity position is big.

In order to quantify the variation of the afterimage pixel, we compare the frames of the group to find the difference of the afterimage pixels. Full the absolute value of the difference of the residual image pixels from f1 to fn, and express the average by M. In eyeball examinations, in order to calculate this with the left and right eyeballs, the value of the right eye is M1 and the value of the left eye is M2.

$$M = \frac{\sum_{i=1}^{n-1} |f_{i+1} - f_i|}{n - 1} \tag{1}$$

Find the sum of M1 and M2. This value is used to verify the relationship between the mean change amount of after image pixels and the degree of fatigue. This indicates that the eye movements are more active as the average variation of the afterimage pixels is larger, indicating that the eye movement is slower as approaching zero.

$$x = M_1 + M_2 \tag{2}$$

Two normal distributions of fatigue state and normal state without fatigue are obtained, and from the difference, the tendency of eye movement of fatigued state is verified. Further, a threshold value of the average variation amount of afterimage pixels is obtained from them and used as a criterion of fatigue degree.

Figure 9 shows the normal distribution obtained in the process of data analysis. The normal distribution of the solid line represents the fatigue state, and the normal distribution of the broken line represents the subject in the normal state without fatigue.

In normal subjects who were not fatigued, values were obtained in the range where the average variation of the afterimage pixels ranged from 0 to 2000, whereas the fatigue state was restricted to a narrow range of about 0 to 800 It is appearing. Also, the vertex position of the probability of normal distribution, the average change amount of the afterimage pixel of the subject in the normal state not fatigued is about 660, but the subject in the fatigue state is about 280, and for the subject in the normal state not fatigued Compared to fatigue subjects, the average change in after image pixels tends

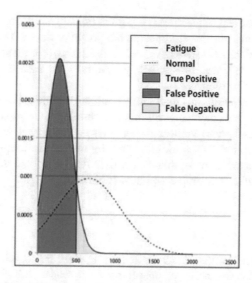

Fig. 9. Normal distribution comparison between normal and fatigue

to be lower clearly. The fact that the mean change amount of the residual image pixels is low means that the eye movements are performed slowly.

The intersection point of the two normal distributions (mean change amount of residual image pixels: 480) is assumed to be the threshold of fatigue degree determination. When the average change amount of the residual image pixels falls below the threshold value, the state is judged as the fatigue state. We analyze the accuracy of this degree of fatigue determination.

The true positivity was expressed in the red area. The true positivity is the probability that a person in a fatigued state is judged as fatigued. On the other hand, the probability that a person in a normal state not fatigued is judged as fatigue, false positives in blue area. The larger the proportion of true positives, the more accurate determination of fatigue level is indicated. The ratio of true positive and false positive is analyzed by ROC curve. In the analysis result this time, AUC = 0.8458, that is, true positivity was about 84.6%. As a result, high degree of independence was found in the results of this verification, suggesting that there is high accuracy for the determination of fatigue level in this study.

6   Conclusion

In this study, we aimed to develop a technique for measuring pixel number variation in gaze direction based on the center of mass of pixels associated with the pupil. The ultimate aim was to establish an objective criterion that physicians can use to corroborate subjective diagnoses of ASD. Using a camera-mounted PC, we developed a high-precision image processing technique to establish an objective criterion that can distinguish individuals with autism from those with normal development.

This measurement technique does not burden the subject; it can obtain data in 15 s. The technique is unique in that it indicates the potential of ASD without the need for observing the individual's behavior. As such, it is effective for identifying children with disorders and will in turn enable the preparation of measures and educational approaches that cater to their needs.

In addition, this research has made it possible to measure the degree of fatigue by developing the eye movement measurement technology we developed so far. This technique can be applied to various fields such as making it useful for preventing dozing driving and improving work efficiency by measuring fatigue level.

We hope that the precision of this tool will be further refined in the future, and that it will be used broadly as a simple and accessible tool.

References

1. Kitazawa, S.: Where tactile stimuli are ordered in time. In: Probabilistic Mechanisms of Learning and Development in Sensorimotor Systems. ESF-EMBO symposia on "Three-Dimensional Sensory and Motor Space", San Feliu de Guixols, Spain, October 2005
2. Kitazawa, S.: Reversal of subjective temporal order due to sensory and motor integrations. In: Attention and Performance XXII. Sensorimotor Foundations of Higher Cognition, Chateau de Pizay, France, 2–8 July 2006
3. Ministry of Health, Labour and Welfare: Guidelines about person of development child with a disability support and the assessment, March 2013
4. Guidelines about person of development child with a disability support and the assessment (2012)
5. Nakano, T., Tanaka, K., Endo, Y., Yamane, Y., Yamamoto, T., Nakano, Y., et al.: A typical gaze patterns in children and adults with autism spectrum disorders dissociated from developmental changes in gaze behaviour. Proc. R. Soc. B. **277**(1696), 2935–2943 (2010)
6. Kitazawa, S.: Eye Tracking Research. http://www.tobii.com/Global/Analysis/Marketing/JapaneseMarketingMaterial/CustomerCases_JP/Tobii_CustomerCase_Clinical_application_and_eye_tracking_gaze_tracking_in_autism.pdf
7. Kimura, Y., Kobayashi, H.: Clinical Study on movement education application of autistic children - approach to enhance the Motor Imitation. The 25th Japan Special Education Society Papers, pp. 446–447
8. Iwanaga, R., Kawasaki, C.: For sensorimotor disorder of high-functioning autism children. Child. Spirit Nerve **36**(4), 27–332
9. Fukuda, S., Okamoto, M., Kato, K., Murata, E., Yamamoto, T., Mohri, I., Taniike, M.: A perception of the others' gaze for children with pervasive developmental disorders. Hum. Dev. Res. **25**, 135–148 (2011)
10. Kawashima, H., Hayashi, N., Ohno, N., Matsuura, Y., Sanada, S.: Comparative study of patient identifications for conventional and portable chest radiographs utilizing ROC analysis. Jpn. Soc. Radiol. Technol. **71**(8), 663–669 (2015)

Measurement of Line-of-Sight Detection Using Pixel Quantity Variation for Autistic Disorder

Takahito Niwa$^{(\boxtimes)}$, Ippei Torii, and Naohiro Ishii

Aichi Institute of Technology, Toyota, Aichi, Japan
b17806bb@aitech.ac.jp

Abstract. In this study, we develop a tool to support physically disabled people's communication and an assessment tool to measure the intelligence index of autistic children, which uses eye movements with image processing. For the measurement of eye movements, we newly developed a pixel center of gravity method that detects in which the direction of the eye movement is shown in the point where the weights of the black pixels moved. This method is different from using the conventional black eye detection or ellipse detection. The method enables accurate detection even when a physically handicapped person uses. On the other hand, the assessment tool that measures the intelligence index of autistic children prepares dedicated goggles that combines light emitting diodes and near-infrared cameras. Further, it measures left and right eye movement differences.

Keywords: Physically handicapped · Autism · Line-of-sight detection

1 Introduction

Physically handicap means having difficulty in activities of daily living, such as walking. Most of the cases, it is caused by a brain disease, such as cerebral palsy. It is difficult for the patients to communicate with family and caregiver. Therefore, in order to know their feelings, various tools have been developed in the past [1, 2]. We have developed the communication tool "Eyetalk" [3] using blinks. It is using the blink detection method using an afterimage, which can detect weak and short blinks of physically handicapped. This application has been widely used not only by physically handicapped people but also people who have become unable to write for communication. However, there is a difference in the conditions of the physically handicapped people. So some people cannot use this application even they need it. It is difficult for the physically handicapped people accompanied by involuntary movements (Abnormal movement of the body that occur independently of the consciousness) to use it. It is necessary to establish a method with new input system.

Based on these experiences and knowledge, we developed a new communication tool that allows input words by the line-of-sight to the screen. Since the developed system has the wide range of input system correspondent to many physical conditions and the wide variety of input technology, the burden of physically handicapped people can be reduced.

© Springer International Publishing AG, part of Springer Nature 2019
I. Czarnowski et al. (Eds.): KES-IDT 2018, SIST 97, pp. 197–206, 2019.
https://doi.org/10.1007/978-3-319-92028-3_21

On the other hand, the autism spectrum is a combination of various categories that have been considered to have a difference in symptoms so far. This is different from the disease of the body with clear symptoms, the symptoms differ for each child. Although the extent of intellectual impairment is defined by IQ, in he/she case, proper communication can not often be obtained even within the usual range.

Since the education and treatment methods at their sites are left to judgment of school and teacher, any indices that can give appropriate guidance to diversity of children are being demanded.

We have developed an assessment tool (objective indicator) to assist subjective diagnosis by doctors. This tool uses dedicated goggles to remove noise by the movement of the head. Further, this indicates that anyone can easily measure weighing auxiliary tools.

2   Purpose of the Study

The number of physically handicapped is the most in the classification of disorders (Visually impaired, hearing impaired, language disorders, limb disorders and internal failure), which is defined in the Welfare Act for Handicapped in Japan. Cerebral palsy, polio (epidemic polio), spina bifida, progressive muscular dystrophy, joint disease, bone disease, brain disorder of movement disorders and a non-brain movement disorders can be cited as symptoms. According to the Cabinet Office Disabilities Report [4] of Japan, the number of the handicapped people who stay at home is 3,864,000 among the total number 3,937,000 in 2011. In other words, about 98% of people with disabilities have been living with family and caregiver supports that are not sufficient enough. Table 1 shows the number of physically handicapped people under and over 18 years old. In recent years, the aging of people with disabilities has become a serious problem. Physically handicapped people who need assistance and care over 65 years old was 31.4% in 1970 and 61.8% in 2006. Also, the autism spectrum integrates various categories such as Asperger syndrome, high-functioning autism, infantile autism, and canner type autism. Unlike physical disorders that often have clear symptoms and causes, each child has different symptoms. The criteria for the extent of intellectual disability defined by WHO are defined intellectual quotient (IQ), 50–69 is mild intellectual disability, 35–49 is moderate intellectual disorder, 20–34 is severe mental retardation, Less than 20 is regarded as the most severe mental retardation. It is difficult to accurately measure these intelligence quotients, and it is required to quantify the intellectual index of autistic spectrum children.

3   Structure of the System

The system changed by blinks was applied in the former study. Manual equipments that allow physically handicapped people who have limitation for movements to carry around have been limited. The picture cards and symbols to determine which direction physically handicapped people looking has been used to communicate with them.

Therefore, we decided to make the application to make decisions by detecting line-of-sight with the image processing.

Haar-like detector of OpenCV is used to detect of the eye area. Around the eye area is darker than the cheek area. Also the upper and lower of the eye areas are cut out in rectangular shapes from the image and the average of the brightness of each area is calculated. When the upper area is bright and the lower area is dark, it is left as a candidate for the face image. In addition, the average of both sides of a nose is often smaller in brightness comparing the top bridge of a nose. Using these criteria reduces millions of face candidate in half. To repeat this method is referred as the high-speed elimination of non-face area by the cascade [5] and it is used in this study.

Many techniques to detect eye have been developed in the past. The Spiral labeling [6] is a blink detection method developed by Tsukada et al. Toyama National College of Maritime Technology. It is a technique for determining the blink when the spiral numbers that assigned to the black area of eyes breaks. However, physically handi-capped people often do not have enough strength to move the eyes. Their eyes cannot be generated spiral because they are half open. So this method did not work. In addition, there are the detection method using the lightness and saturation of the eyes [7] and the complexity [8] to capture the outlines of eyes and determine the closing of eyes when it becomes flat. These methods are effective for a healthy person. But it was difficult to capture the weak and short blink of a physically handicapped person.

Therefore, we have developed a blink detection method using the afterimage [9]. This is a new method to use the point where the pixels of black area has not moved for a certain period of time as the base by recording the pixels of positions of the black area of eyes. This can be responded to weak and quick blinks. Afterimage records the position of the black pixel of the eye unit. All of the pixels have a box of variable called Burn-in value. Pixel which are determined to black adds 5 to Burn-in value, and it subtracts 1. Further, the determination pixel is not a black subtracts 1 (Not that the value is less than 0). When Burn-in value has exceeded the 50, that can be used for blink determination as an afterimage (Fig. 1).

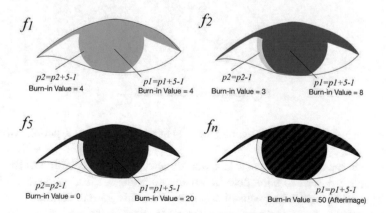

Fig. 1. The image of afterimage method

Equation (1) shows the value of added to the Burn-in value for the determined pixel state *(p)*. And Eq. (2) is a method of obtaining the Burn-in Value *(V)* of the pixel at the coordinate *xy*.

$$a_f = p - 1 \tag{1}$$

$$V_{xy} = \sum_{f=1}^{n} a_f \tag{2}$$

Only if the user performs a series of operations of the closed-eye and eye opening from eye opening was evaluated as "blink". Replacing this as a change of the pixel, reduced at once by blinking the number of pixels defined as an afterimage, it may be regarded as only when subsequently recovered.

The advantage of this afterimage technique is that it can be judged a quick blink and weak blink, It's of course, It's like opening the degree of brightness and body movement and eye of the room, less likely to be influenced by the physical characteristics and external factors of the user.

For line-of-sight detection, there are ones using circle and ellipse detection of black eye. However, even for healthy people, unless eyes are wide open the eyes can not be detected (Fig. 2).

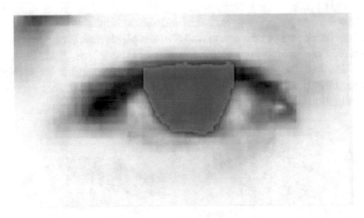

Fig. 2. The shape of black eye

These techniques are not suitable for physically handicapped people who often have half-open eyes (Fig. 3).

Therefore, we developed a new function to detect the gaze direction of the user by using the above-described blink detection using afterimage. First, we acquire pixels of black eye in afterimage, and create criteria for detection of gaze direction. Compare this center of gravity to the left or right in the current frame with the afterimage and use it to detect the gaze direction. The center of gravity is the pixel with the highest numerical

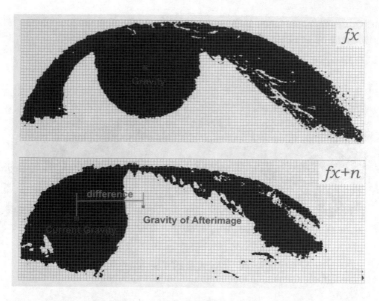

Fig. 3. Comparison of afterimage and current frame

value when dividing the number of pixels determined to be black pixels from all Burn-in Value values.

Also, as in the series of motion of blinking, the line of sight direction was defined as the "movement of the intended line of sight" when the line of sight moved and returned to the original position again. Equation shows the motion difference ΔT between the current frame (f) for gaze direction detection and the central pixel (g) one frame before $(f - 1)$.

$$\Delta T_g = \sum_{f=1}^{n} a_f - \sum_{f=1}^{n} a_{f-1} \tag{3}$$

In general, the center of gravity refers to the point where the primary moment of the surroundings is 0, that is the point where the weights are balanced.

Thus, the center of gravity is obtained.

Pixel center of gravity is difficult to grasp the center of pupil accurately. However, processing speed is much faster than circle and ellipse detection. And also, with the line of sight facing sideways, even if the shape of the black eyes collapses, the center of gravity can be accurately grasped.

After defining the center of gravity of the pixel, the movement of the black eye is detected. The center of gravity of the afterimage is compared with the center of gravity of the current frame, and when there is a difference more than a certain amount, it is determined that the black eye has moved (Fig. 3).

Next, the line-of-sight direction is detected. When the eye part image is cut out, pixel coordinates in the x axis and the y axis are generated from the upper left to the lower right of the image. The center of gravity of the afterimage is compared with the x coordinate of the current frame. If the coordinate is small, it is detected to the left, if it is large, movement of the line of sight is detected to the right (Fig. 4).

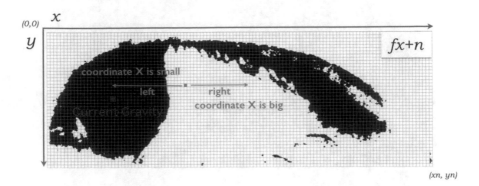

Fig. 4. Judgment of line-of-sight

4 Application to Assessment Tool for Autism

Based on the research and development so far, the amount of change in the number of pixels of both eyes is obtained, the severity of the autistic group is judged from the deviation of the left and right eye movements and the reaction speed, the objective index (threshold) is determined therefrom.

For gaze measurement, use goggles to blink four light emitting diodes in the up, down, left, and right in order, and make the direction of light gaze with line of sight. Capture left and right eye movement differences from the near infrared camera by pixel number. Figure 5 shows the developed eye movement measurement goggles.

Fig. 5. The eye movement measurement goggles

Near infrared cameras are installed on the left and right inside the goggles. Moreover, Raspberry Pi which is a compact single board computer equipped with an ARM processor is built in, and work of synchronizing the light timing of the eye movement measurement program and the light emitting diode is carried out. Figure 6 shows the internal structure of the goggles.

Fig. 6. Internal structure of the goggles

Using the measurement result of the detected eye portion data, verify the relationship between the eye movement and the intelligence index.

In an everyday visual environment, when an object of interest is moving in the field of vision, two kinds of eye movements occur to see the object. One is a fast, transient eye movement that directs the line of sight to the object, called saccade. This eye movement acts to capture the image of the object in the fovea of the retina. The other is a slow and smooth eye movement that works to keep the image of the object in the retina central fossa and is called tracking eye movement (passat) [9]. The author assumed that the sensation became dull and the speed of directing the line of sight to the object was reduced, that is, the saccade was abnormal. In this study, we display saccadic tasks on developed goggles and detect eye movements of subjects. Contents of the prepared saccade task lights the light emitting diodes arranged vertically and horizontally randomly and measures the reaction time until the eyeball starts to move.

Analyze based on data obtained by eye movement detection.

In step 1, in order to perform afterimage analysis, the afterimage pixel (the position of pixel centroid) of the detected eye movement is stored for each frame. Here, the afterimage pixel is f, the first frame is $f1$, the second frame is $f2$, and the latest frame is fn.

In step 2, in order to quantify the amount of change of the afterimage pixel and eye movement, the difference between the afterimage pixels is obtained as an absolute value by comparing the preceding and succeeding frames. The absolute value of the difference between the residual image pixels from $f1$ to fn is obtained, and the average value is indicated by M. In eyeball movement detection, the right eye value is set to $M1$ and the left eye value to $M2$ in order to calculate with the left and right eyeballs.

$$M = \frac{\sum_{i=1}^{n-1} |f_{i+1} - f_i|}{n-1} \tag{4}$$

In step 3, the absolute value of the difference between $M1$ and $M2$ is obtained and is taken as x. Let this value be the average change amount of the afterimage pixel, and the smaller the average change amount of the afterimage pixel, the more the left and right eye movements are equal. As the numerical value increases, the deviation of the eye movement increases.

$$x = |M_1 - M_2| \tag{5}$$

In step 4, the standard deviation of the mean change amount of each afterimage pixel is obtained by dividing it into subjects of autistic children and healthy subjects. The standard deviation is denoted by s, and n is the number of subjects with autistic children or healthy subjects.

$$S = \sqrt{\frac{1}{n} \sum_{i=1}^{n} (x_i - \bar{x})^2} \tag{6}$$

In step 5, a normal distribution is expressed by a probability density function using the standard deviation obtained in step 4. The equation for finding the normal distribution is $f(X)$.

$$f(x) = \frac{1}{\sqrt{2\pi s}} exp\left(-\frac{(X - \mu)^2}{2s^2}\right) \tag{7}$$

We obtain two normal distributions of autistic children and healthy subjects and verify the tendency of eye movement from the difference. From the two normal distributions, a threshold value of the average change amount of afterimage pixels is obtained and used as a criterion for judgment.

5   Toward Practical Application for Measurement

The eye movement detection method and the data processing method can be applied by applying the previous research. However, the problem still remains in the image processing method due to the use of the infrared camera. There is not much difference between the brightness of the eye portion image output from the camera and the brightness of the irises and there is a point that it is difficult to obtain the accurate position of the black eye when binarizing. Figure 7 left side is the acquired eye image, and the right is the binarized processed image.

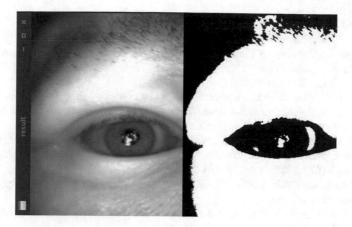

Fig. 7. The saccade program

In the future, we will re-examine the method of processing the eye image (adjust contrast, brightness/saturation, binarization, etc.) and make improvements so that appropriate gaze direction detection can be performed.

6 Conclusion

In this study, we developed a communication support tool for physically handicapped persons and an assessment tool to measure the intelligence index of autistic children using eye movement measurement by image processing. For the eye movement measurement, the pixel centroid method was used, which detects in which direction the point at which the weights of black pixels moved, without using the conventional circle detection of a black eye or ellipse detection. This makes it possible to detect accurately even when a physically handicapped person uses it. The developed program has already been released as "Eyetalk Pro" [10], and the use is spreading mainly by special support schools.

The assessment tool to measure the intelligence index of autistic children developed dedicated goggles that combine light emitting diodes and near infrared cameras, developed a system to measure left and right eye movement differences, and now improve accuracy for practical use.

In the past, our research was not a methodology, but actually applied the developed equipments, investigated how it was utilized, and repeatedly applied it. Also in this research, I would like to listen to the voice of users, find problems and make use of it in future improvements.

References

1. Kitazawa, S.: Reversal of subjective temporal order due to eye and hand movements. In: ESF-EMBO symposia on "Three-Dimensional Sensory and Motor Space", SantFeliu de Guixols, Spain, 10 October 2007
2. Kitazawa, S., Nishida, S.: Adaptive anomalies in conscious time perception. In: Tutorial Workshop in the 12th Annual Meeting of the Association for the Scientific Study of Consciousness, Taipei, Taiwan (2008)
3. Torii, I., Ohtani, K., Niwa, T., Ishii, N.: Detecting eye-direction using afterimage and ocular movement. In: ACIS 2nd International Symposium on Computational Science and Intelligence (CSI 2015), pp. 149–154 (2015)
4. According to the Cabinet Office Disabilities Report (2015). http://www8.cao.go.jp/shougai/whitepaper/index-w.html
5. Lienhart, R., Maydt, J.: An extended set of Haar-like features for rapid object detection. In: IEEE ICIP 2002, vol. 1, pp. 900–903 (2002)
6. Tsukada, A.: Automatic detection of eye blinks using spiral labeling. In: Symposium on Sensing via Image Information (SSI 2003), vol. 9, pp. 501–506, June 2003
7. Torii, I., Ohtani, K., Niwa, T., Ishii, N.: Study and development of support tool with blinks for physically handicapped children. In: 2013 IEEE 25th International Conference on Tools with Artificial Intelligence, Herndon, VA, USA, 4–6 November 2013, pp. 116–122. IEEE Computer Society (2013)
8. Torii, I., Ohtani, K., Niwa, T., Ishii, N.: Development of support applications for elderly and handicapped people with ICT infrastructure. In: HCI International 2013 - Posters' Extended Abstracts - International Conference, HCI International 2013, Las Vegas, NV, USA, 21–26 July 2013, Proceedings, Part I. Communications in Computer and Information Science, vol. 373, pp. 266–270. Springer (2013)
9. Torii, I., Ohtani, K., Ishii, N.: Study and application of detecting blinks for communication assistant tool. In: 18th International Conference in Knowledge Based and Intelligent Information and Engineering Systems, KES 2014, Gdynia, Poland, 15–17 September 2014. Procedia Computer Science, vol. 35, pp. 1672–1681. Elsevier (2014)
10. Eyetalk Pro (Apple App Store). https://itunes.apple.com/jp/app/aitokupro/id1073335134?mt=8&ign-mpt=uo%3D4

Irradiation of the Ear with Light of LED for Self-awakening

Mateus Yudi Ogikubo$^{(\boxtimes)}$, Ippei Torii, and Naohiro Ishii

Aichi Institute of Technology, Toyota, Aichi, Japan
x17025xx@aitech.ac.jp

Abstract. In recent years the people are more and more exposed to blue light by increasing in desk work. Since the blue light has awakening effects to disturb sleep, so it tends to cause lack of sleep. Therefore, strong sleepness is more likely to appear during the daytime. A nap is recommended as a method to prevent sleepiness. Also, there is a research, which shows that self–awakening is effective for sleepiness. We consider it can promote self-awakening to irradiate the ear canal with blue light to make a nap effective. In order to prove this, we developed a device that can irradiate the ear canal with LED light. As a result, we confirmed the increasing of brain blood level and changes of brain waves by illuminating the ear canal with blue LED light. From these experiments, using light can promote self-awakening. Even a person without the custom of self-awakening can also be awaken by himself. Sound and vibration will be effective in order to realize the awakening.

Keywords: Self-awakening · Brain blood volume · Brain waves

1 Introduction

A large kind of the animals live a same life rhythm, by which they get up with the sunrise, they will be active during a day and they will take a rest with the sunset. The human being lived the similar life rhythm before. However, people started to active at night by using the light of the fire since they became able to use the fire. As a result, people have more time to active and less time to sleep. we paid attention to the research which was carried out in Oulu University that an antidepressant effect appeared in a light therapy of the light in white color irradiating to the ear canal to deal with seasonal affective disorder [7]. In this research, the palliation of depression symptom was found in approximately 70% of the subjects who received SAD diagnosis. Based on this research, Valkee company developed a bright light headset as an apparatus for therapy to fix the life rhythm [8]. From this, we thought that brain has reaction to light by irradiating light to ear canal and decided to irradiate light to an ear instead of eyes. Based on these, we carry on the research of developing the device which promoted self-awakening by irradiating ear canal with light. The measurement location was performed in a quiet, dark room. We asked 23 subjects without a custom of the self-awakening took a nap when we measured it to get the data. We asked the subjects to take the experiment in the environment of "a nap without the light" and "the nap that there was light" to compare the effects of the light in the nap. In this research, we used

© Springer International Publishing AG, part of Springer Nature 2019
I. Czarnowski et al. (Eds.): KES-IDT 2018, SIST 97, pp. 207–216, 2019.
https://doi.org/10.1007/978-3-319-92028-3_22

the near-infrared spectroscopy (NIRS) to measure the brain blood volume, and elec-troencephalograph (Brain Pro) to measure the brain wave.

2 Explanation of the Developed Device

2.1 Prototype 1

In this research, to control a nap by irradiating ear canal with light, we developed an earphone-shaped emission of light tool (Fig. 1). In the development of the emission of light tool, we adopted the high-power LED with high luminous efficiency, and rela-tively small calorific value. In addition, we decided to use a volume transformation device for a converter from the viewpoint of ease and robustness of the handling. In this way, the prototype became a small device with easy method to handle (Fig. 2).

However, folding this device with high-power LED needs big electricity that there is much consumption of the dry cell. And we changed the resistance of the prototype so that the heat would not affect the measurement result when the light turned strong. It was a problem that the light could not get through the ear canal because the size of the high-power LED was large.

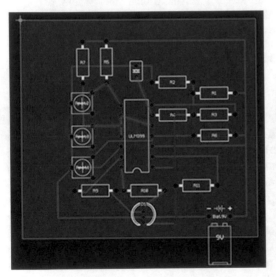

Fig. 1. Light emitting device design circuit surface

Fig. 2. Volume device

2.2 Prototype 2

We developed prototype 2 with Arduino to improve problems of prototype 1. Arduino is a computing system, a board with a small computer put on it and using software to develop a program and a kind of programming language [9].

We connected this device to a PC and enabled to adjust the illumination brightness of the LED light by writing the program. In addition, it was able to solve the problem of prototype 1 with consumption of the dry cell because electric supply was performed from a PC (Fig. 3). And we changed the high-power LED into canal type high brightness LED to solve the problems that was change the resistance to heat did not affect the measurement result when the light turned strong, the light could not get through the ear canal because the size of the high-power LED was big. Using the prototype 2 we became able to measure data more stable than prototype 1. In addition, prototype 2 using the program became able to flash an LED light which was regular lighting in prototype 1. Therefore, we came to be able to compare the regular lighting with the flashing lighting to distinguish which was more effective.

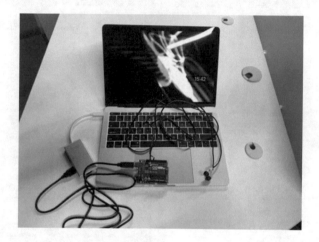

Fig. 3. General view of the system

3   The Measurement of the Influence by the Light

3.1   Brain Blood Volume

NIRS is a device that measure the blood volume in the brain noninvasively and visualizing a blood volume change of the frontal lobe. The hemoglobin of the blood component lets light scatter, and the degree of the dispersion changes when oxygen connect to this. The absorption of the light in the wavelength area of near-infrared light occurs by oxygenation hemoglobin (oxyHb) and deoxygenation hemoglobin (deoxyHb), and both have different absorption spectrums. And if a molar molecular extinction coefficient of oxyHb and deoxyHb is known, it is able to calculate the density changing of oxyHb and deoxyHb by measuring the absorbance changes in higher than 2 wavelengths, at the same time the graph of the brain blood volume would be displayed in real time on the desktop (Fig. 6). Then the data would be output as text data after a measurement (Figs. 4 and 5).

Fig. 4. Near-infrared spectroscopy (NIRS)

Fig. 5. Wearing the NRS

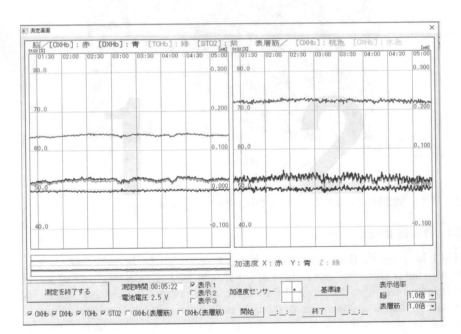

Fig. 6. Measurement screen of NIRS

We paid attention to relations between oxyHb and deoxyHb to confirm a nap state of the subjects. When people sleep they become relax, the quantity of oxygen supplied to the brain decrease and becomes constant later. We defined the state that the amount of oxyHb was greatly less than deoxyHb and keep constant as falling asleep while using the amount of deoxyHb as a judgment of falling asleep and the awakening (Fig. 7). In addition, even if there is more oxyHb than deoxHb, it not means that the subject is awaken. We judged the awakening of the subject from self-report and a brain blood volume.

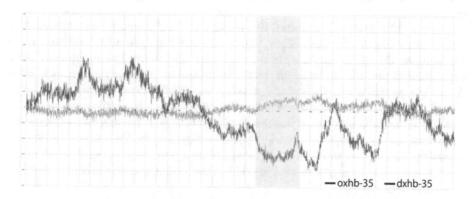

—oxhb-35 —dxhb-35

Fig. 7. The yellow part is the time when the subject fell into the asleep

We expected that the blue light with awakening action raised the amount of oxyHb and supposed that we could control the timing of the awakening by controlling the increase and decrease level of the oxyHb using the LED and tested it. We made the subjects to sleep attaching the NIRS and prototype 1 to confirm that light of prototype 1 arrived in brain through ear canal. And to avoid the placebo effect, we did not tell the timing of lighting to the subjects. After having confirmed the subjects slept by analysing real-time data from NIRS, we turn on the blue light of prototype 1 and observed

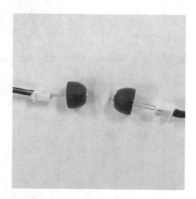

Fig. 8. Large high-power LED

Fig. 9. Canal type high brightness LED

the changes of the brain blood volume. However, the brain was unstable in a reaction to prototype 1 which focused on the quantity of light. We could not observe a stable change because light did not go through ear canal due to the part with light of prototype 1 was large (Fig. 8). We changed a lighting part of prototype 1 into canal type high brightness LED to solve it, and made the light going through surely to the ear canal (Fig. 9).

As Fig. 10 shows, we confirmed that the oxyHb level rose after having turned on the blue-light LED during sleep. From Fig. 10, we can find the oxyHb level often made the rising on the right, but there are data that after oxyHb level going up it will turn stagnated or decreased. It is thought that the LED light of the prototype arrives in brain due to the oxyHb levels rose after irradiating the light. However, the subjects who cannot fall asleep were not able to confirm the same change even if we carried out the same experiment. We supposed that the oxyHb level repeats the increase and decrease when the human beings are awaking because they are thinking about something. Therefore, as it shows in Fig. 11, we thought that it was not possible to confirm the same change during a nap to an awaking subject even we irradiated the ear canal of him with LED light.

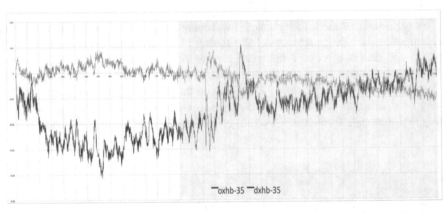

Fig. 10. The yellow part is the time when the LED of prototype 2 is turned on

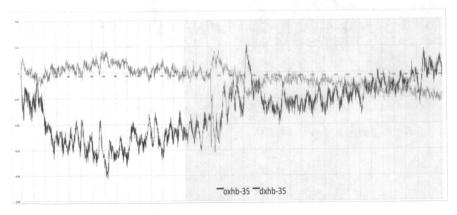

Fig. 11. The yellow part is the time when the LED is turned on

Then, it is necessary to check how long the it takes for the brain to react to the light go through ear canal after the light turning on. Because if it takes a long time until brain react to light, it is not able to use the light for the assistance to nap for a short time. We got the data of average time from irradiating the blue light in both ears of the subject to the brain blood volume and brain waves changing. We defined the oxyHb level of the time we turned on the light as the standard value, and the time of making the oxyHb higher than 0.01 of standard value as react time. We calculated the average time by calculating the lag time of brain reacting when we turned on the light. As the result, the average time was 36.6 s.

There are data that the oxyHb level which rose by a reaction of the light decreased later. Then, we thought that the oxyHb level could continue rise using the period of the timing to turn on the light. Because the prototype 1 could not flash in constant period, we developed the prototype 2 that is able to control the flash by program.

We looked for the most effective lighting period and pattern while regulating a lighting period and a pattern by a program. In pattern 1, we set lighting time as a cycle that the lights turned on and off for 0.07 s. However, the oxyHb level did not have a big change in pattern 1. We thought that the main cause would be the extremely short time of lighting so we decided to delay the period of lighting. We set it to replace lights out to lighting time with pattern 2 every two minutes. In result, when we irradiated ear canal with pattern 2 after falling asleep, we confirmed the oxyHb level rose generally and there was not big decrease found in Fig. 12. From this, it can be thought that two minutes period is more effective than regular lighting for the increase in oxyHb level.

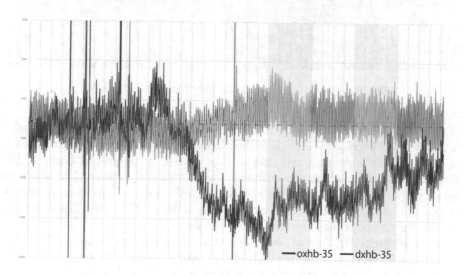

Fig. 12. The yellow part is the time when the LED is turned on with pattern 2

3.2 Brain Waves

Brain Pro is the device which can measure brain wave by attaching a sensor band to frontlet. Generally, it is unsuitable for the electroencephalographic detection in the frontlet because it is easy to make noise by a blink or the ocular exercise. However, when a noise occurred, a function of Brain Pro prevents the noise from affecting the result of a measurement by stopping the measurement so this electroencephalograph enables the brain waves measurement from frontlet. And we can confirm the numerical value of 3.0 Hz–30 Hz displayed in real time by a desktop with optional software (Pullax F). We can grasp the state of the sleepiness and relaxation degree of the subject by using these. In addition, it prevents the defeat of data by sampling the high density of 1,024 points in one second at the time of AD conversion. The measured data are saved in CSV form after a measurement [11] (Fig. 13).

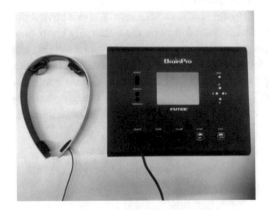 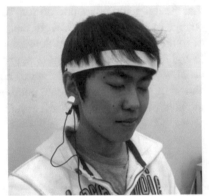

Fig. 13. Change in cerebral blood flow/Wearing the Brain Pro

We converted data obtained from Brain Pro into an alpha wave, a beta wave and a theta wave and observed the changes in the brain. The alpha wave of the person rises at the time of relaxation state and decreases in inverse proportion to depth of the sleep. The beta wave of the person rises at the time of strain state and decreases in inverse proportion to depth of the sleep as well as an alpha wave. The theta wave shows a sleepiness and sleep state, it rises in proportion to depth of the sleep and is used to measure the degree of the sleep inertia after the sleep.

In the research of Institute of Nanjing physical education, they divided subjects into 60 min and 90 min groups at nap time and checked the relations between sleep and brain waves. As the result, the brain waves of each group got the similar increase and decrease. After the subjects slept, an alpha wave and the beta wave suddenly became higher and lowered as sleep was deepened. Then it became higher again as it is close to the time of awakening. On the other hand, theta wave and delta wave moved reverse to alpha and beta wave [12].

We based on these findings and used it for judgment if the light irradiated ear canal is effective in a nap. We changed the program of prototype 2 to keep it always on, and after the subjects slept, we turned on LED in both ears and observed the brain waves state. As the result, the alpha wave increased after we turning on the blue-light LED. From a research of Institute of Nanjing physical education, we thought that we could awaken the subject who was sleeping by irradiating ear canal with light because it is thought that alpha wave and beta wave are high so that sleep is light. In addition, it is thought that the brain waves would be affected by the light up because we can find the brain becomes active (Fig. 14).

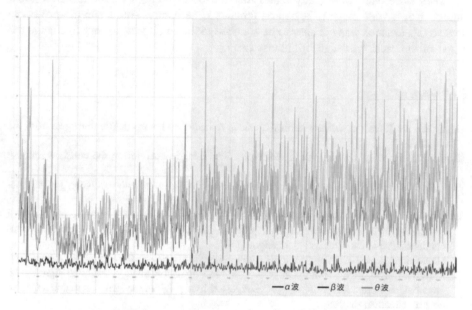

Fig. 14. The yellow part shows the time when the LED turns on

4 Conclusion

From the results of this research we could confirm a thing that the brain waves and oxyHb levels increased by irradiating the ear canal with light of the blue-light LED during sleep. From this, it was revealed that the brain could carry on the awakening action by perceiving the blue light from an ear and that we could induce a person to be awaken during sleep by using blue light. In addition, the time for brain to perceive light and react is short, so we can use it for a nap. Furthermore, we were able to confirm it was more effective than lighting all the time by setting the cycle of lighting time and lighting out time to two minutes.

Also, from the research result, we think that the person who cannot self-awake, elderly person and a person with the heart disease can be awakened in a state similar to self-awakening without any burden on the heart by irradiating an ear with blue light. In addition, we have controlled sleep inertia by taking the nap in this way with the light, thus we can use it to improve the work efficiency and concentration during the daytime.

4.1 Future Development and Research

In this experiment, we aimed to use blue-light LED to make a nap with less sleep inertia. The next step is to introduce low-frequency and sedating music at the time of arousal, and to develop research to increase the effect of reducing sleep inertia in order to further improve the effects of sleep for a short time. Also, for the light used for promoting sleep and promotion of arousal, we have not considered the influence due to the differences in light intensity, comparison in different colors and so forth, so we will still carry on research including these.

Also, since there were many males in their teens and 20 s in this research, measurement by a wide range of ages and females is necessary. And since the prototype used in this research was not easy to be attached to the ear, it is necessary to change the shape of the canal type high brightness LED.

References

1. Kaida, K.: Effects of self-awakening on sleep structure and sleep inertia during or after the short nap
2. Carskadon, M.A., Dement, W.C.: Multiple sleep latency tests during the constant routine: Sleep 15 (1992)
3. OECD Balancing paid work, unpaid work and leisure. http://www.oecd.org/gender/data/balancingpaidworkunpaidworkandleisure.htm
4. Ministry of Health, Labour and Welfare. Summary of nation health and the nourishment findings (2015). http://www.mhlw.go.jp/file/04-Houdouhappyou-10904750-Kenkoukyoku-Gantaisakukenkouzoushinka/kekkagaiyou.pdf
5. Hayashi, M., Hori, T.: A short nap as a countermeasure against afternoon sleepiness (2007)
6. Masahiko Ayaki, Ch.: Biological effects of blue light contained in artificial lighting on circadian clock and sleep/awake cycle-suggested living lighting system for maintenance of healthy circadian rhythm
7. Oulu University Transcranial bright light treatment via the ear canals in seasonal affective disorder: a randomized, double-blind dose-response research. https://www.ncbi.nlm.nih.gov/pmc/articles/PMC4207317/
8. Human Charger. https://humancharger.com
9. Arduino. http://www.ric.co.jp/book/contents/pdfs/879_p018.pdf
10. Shimadzu. http://www.an.shimadzu.co.jp/bi
11. Futek. http://futek.jp/products/brain/fm_929/index.html
12. Journal of Nanjing Institute of Physical Education. Research on EFG-based Noon Staging for Athletes (2013)

Projection Mapping of Animation by Object Recognition

Kazuya Yonemoto[✉], Ippei Torii, and Naohiro Ishii

Aichi Institute of Technology, Toyota, Aichi, Japan
x16101xx@aitech.ac.jp

Abstract. Development of projection mapping in recent years has remarkable attention, being used at various events for advertisements. Though being used, its production-pattern is constant with fading steps and novelty. Thus, we thought of a new projection mapping system instead of an 'only-seeing' projection mapping system. We decided to develop a way in which the picture changes according to the shape of the mapping. We can specify the size-ranges to acquire a more precise shape. Then, we can develop an application (called "Video switching") that passes the image which suits its reference data among collections of the data. A coordinate of an object is also included in the received data, and they become able to reflect a picture in a coordinate of an object. A number is assigned to label the recognized object. It distinguishes the combination of an object by this number. Moreover, it enables to distinguish how many groups of an object exist by this number. This research develops the visualization in such a way that changes in combination and are projected comfortably. Also, this research focuses on increasing the formation of an object which can be recognized and projected.

Keywords: Projection mapping · CG · Object recognition · Real-time property

1 Introduction

In the projection mapping we have created so far, the image changes according to the picture by the projection mapping. As the image changes in response to its sound, people are delighted to watch it. In addition to listening and entertaining, we intended to create something that can be seen and experienced by being interactive. Therefore, we thought that if we could not only look, listen and entertain but also interact with the work to experience them.

Projection mapping is a picture-technique that projects computer-graphic (CG) images onto objects and spaces. It is an illusionary image-technique that makes the color and shape of the object projecting the image look as if they were changed. However, with the improvement of technology in CG, image representation which is increased by the high quality is used in many areas. Initially, the projection mapping's curiosity attracted a lot of attention. With it spreading all over the world, it got common and now anybody can seek for a new style, and there is the necessity to invent new techniques and ideas to respond to the production.

© Springer International Publishing AG, part of Springer Nature 2019
I. Czarnowski et al. (Eds.): KES-IDT 2018, SIST 97, pp. 217–226, 2019.
https://doi.org/10.1007/978-3-319-92028-3_23

In the projection mapping we have created so far, we decided the place to project first and then produced and projected movements and effects that fit the determined sound. However, when collaborating in a live performance at a certain event, the player had to play in accordance with the musical notes-mapping. This causes a feeling of incompatibility when the scale deviates from the performance of the performer. We realized that it would be a great challenge to completely eliminate the difference between "expression" and "production" like this. Therefore, in order to solve this problem, we worked on creation of contents in which the image projected according to the shape would change. The produced content recognizes the shape formed by changing the combination of the blocks and the video matching the shape is instantaneously displayed. Thus, the real-time property is never lost, however the coordinates acquired when the object is recognized are not stable. Therefore, we try to stabilize the coordinates to be acquired.

2   Purpose of the Research

This research aims to have an interactive nature to develop a new expression method by projection mapping. This is to make people feel more familiar with projection mapping, which until now was just enjoying through seeing and experiencing.

There are existing ones which can actually be experienced by people moving by themselves. However, as we move the body, we cannot experience it unless we manage to have a wide area. Thus, we thought that this interactive projection mapping could be experience without securing a large area or the exercise. We thought it would be unnecessary to secure a wide place if it could be experienced without exercise, and cost could be reduced in terms of equipment installation. Therefore, we started research with the aim of acquiring the shape of the block in the stereoscopic space using Kinect (motion sensing device) and implementing the technique of projecting the image at the same position as of the place where the shape was acquired.

By recognizing the shape of the block in the stereoscopic space, it becomes possible to change the mapping image being projected according to the changed combination, even if, the combination of the projecting blocks is changed. With this, not only the blocks that are projecting mappings to the spectators watching the projection mapping can be experienced, but also the interactivity as they actually participate by touching and moving. However, this alone only changes the shape of the block that can be combined, the image being projected does not change. Thus, we decided to install a technology to project the video at the same position of the projecting block so that the video will switch according to the deformed shape and this time we will have a story property. By giving a story property, since images are switched according to the change of the shape made by the recognized combination, it is possible to implement real time interactive projection technique by mapping without losing property in real-time.

Furthermore, in order to feel more experienced that the image changes, we made it to automatically switch images according to changing combinations. When the combination and the coordinate point are always acquired and matched with the determined shape, the image being projected is switched to the image corresponding to the position of the combination formed by the combination and displayed. The image to be projected has a story property so that people who are experiencing do not get bored.

Moreover, it is possible to substitute not only a mapping image on the shape formed by combining it as a novelty of contents but also a familiar object without using a dedicated block for displaying the expansion of the story. Since, it can be used even for things, it aims for high versatility.

At the end of this development, we will be able to recognize not only blocks but also complex shapes such as cylinders, triangular prisms, spheres, etc. that can recognize shapes, not only flowing images when they are recognized but also different sounds depending on recognized objects. It would make sounds when moving the object so that the color changes according to the recognized object and the trajectory remains after moving the object and the color remains. To improve, we need to implement these ways so as to expand.

3 Structure of the System

In this research, in order to perform interactive mapping, the application (henceforth referred to as Box Detector), Kinect (to recognize the form of motion-processing), flash animation, and an object are used. Interactive mapping processing is an Integrated Development Environment which specializes in different areas, such as art and design, and can simplify visual expressions. This processing was easy to use for a student, an artist, and a designer and also suitable for an interactive visual expression. Thus, we infer that it is the best to use for this research.

In this processing, it is decided to processes the data of number of the object (Fig. 1) acquired by Box Detector, coordinates, and the combined form. Then, it passes a flash animation according to the acquired data, and expresses the interactive mapping.

Fig. 1. The block to project (cube of the styrene form whose one side is 100 mm)

Fig. 2. A Kinect set

3.1   Object Recognition by Kinect

A high-definition depth-sensor and RGB camera were highly precise than Kinect in the past. Until now, Kinect needed to be placed in a higher position (height-wise) and farther (distance-wise). However, 'Xbox One' Kinect was used in this research. This is because Xbox One Kinect recognizes the objects accurately which improves the macro (short distance) view as well as the majestic (tall height) view. Thus, it can be used in a small-scale environment (such as a small room) as well with precision. Furthermore, active Infra-Red (IR) rays is adopted for darkness-issues, such that when a strong light hits the object from one side a shadow occurs on the other, IR rays detect and recognize the light which is not influenced by visible light. Moreover, it's wide-angle uses 60% from the old Kinect and the recognition range spreads changing the principle of 3D-sensor. Both of the resolution and the depth-direction is strengthened in an RGB camera. It is changed from VGA (640 × 480) to 1080 pixels. Therefore, it decided to use Xbox One Kinect (Fig. 2).

3.2   Flow Chart

The flow of the processing currently performed processor, as shown in Fig. 3, is as follows. First, the measurement range of Kinect is set. The upper left of a screen and a lower right coordinate-values are acquired so that it fits the range of projector. Next, a block is put in the area and a virtual floor to the object is set. This makes the setup of XY coordinates easier within processing. This virtual floor can be set manually. It is also possible to acquire the shape of the object with higher accuracy. Moreover, the acquisition range of either sides can also be set. When recognition is started, a wall may be recognized as a block. In order to avoid this, it is possible to set the color to be recognized. Until the setup of range is carried out, an object is automatically detected by Box Detector to some extent.

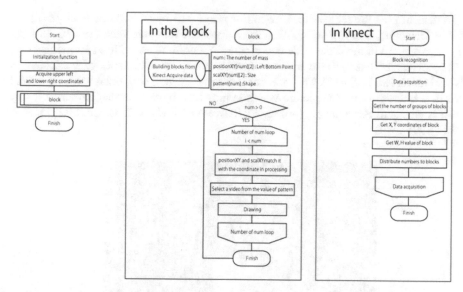

Fig. 3. Flow chart of processing by object recognition

3.3 Object Recognition Program

The location doesn't fit anymore because the standard is different from a coordinate in processing by the coordinate which is still acquired from Kinect. Thus, calibration is done. Calibration originally means 'to adjust the instrument scale correctly' or 'to adjust the electronic circuit to comply with standards and standards'. However, it is used also for changing the information acquired by the sensor according to the purpose-of-use. Firstly, as a procedure of calibration, it changes into rectangular coordinates from a polar coordinate. It becomes easier to use it by performing this for the function in processing. However, since the inclination of size, the coordinates acquired by the sensor and the coordinates inside processing as the starting point, and an axis etc. are different, an adjustment is added. In order to rectify deviation, it describes the maximum and the minimum of XY of the range projected by a projector, and Processing. In search of a homograph procession, a projective transformation is performed using the coordinates of four corners of screen-size. As a result, the reference of coordinates is corrected and becomes usable coordinate-data. Block is recognized by Kinect based on this coordinate-data. The virtual floor is set to the object by flow height (FH), value of which can be set up manually. In order to perform more accurate block-shape recognition, it is possible to reduce the extra measurement range by adding FH to the base in the block. It isn't only FH that it can be set. By recognizing the block by changing the setting of threshold (TH) it is possible to remove what is recognized other than the block. Then, Max Depth (MD) sets the distance between Kinect and the object and improves the accuracy of recognition. Moreover, the acquisition-range on either side can also be set up by left padding and right padding with (LW and RW). The size of an object to recognize as a block can be set up by changing the value of box size (BS).

After finishing the setup of the recognition range, an object is automatically detected by Box Detector (as shown in Fig. 4). If it is a result of Fig. 4, the values of 216, 230, 50, 47 are expressed in the order of X coordinates, Y coordinates, width, and height.

An object is detected and also the number of every coordinate of the object is divided (as shown in Fig. 5). This number is assigned with a horizontal value $(0\sim)$ and a vertical value $(0\sim)$). The picture which is indicated by the number to which it was given is changed. For example (in Fig. 6), the value 1011021222 is acquired if it is A and is 0001 for B.

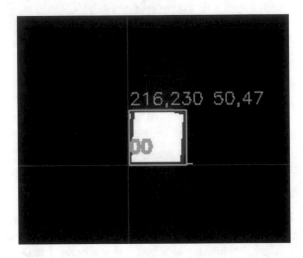

Fig. 4. The figure and numerical value which were detected

(0,0)	(1,0)	(2,0)	(3,0)
(0,1)	(1,1)	(2,1)	(3,1)
(0,2)	(1,2)	(2,2)	(3,2)
(0,3)	(1,3)	(2,3)	(3,3)

Fig. 5. The number assigned by coordinates

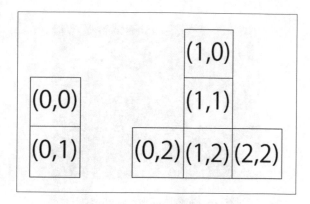

Fig. 6. The detected figure

3.4   Image

The image manufactured this time is divided into three steps, and the image projected with the combination of block changes. A child as displayed in Fig. 7, is within a block. The child can move within the limits of a block. The picture from which a soybean tree grows when shipping another block lengthwise placing three blocks vertically, the bean tree grows bigger and the child climbs the bean tree. Thus, a story shall develop as a block is newly added (Fig. 7).

Fig. 7. The tree of the child who appears in a story, and beans

3.5   Mapping Projected onto a Block

The picture made this time is projected in a block. As there were various kinds of projectors, the projector with the high resolution suitable for a mapping was selected this time. In this research, we decided to put it just beside the projector to make sure that the location of the projector and Kinect won't bother the mutual functioning (as shown in Figs. 8 and 9).

Fig. 8. Placement of the projector

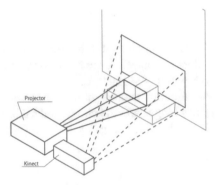

Fig. 9. Simplified drawing of the acquisition range of Kinect and the projection range of the projector

4   Conclusion

- Display of video according to the shape of the object

The image to be displayed is divided into numbers formed by combining blocks one by one, and the number assigned by Kinect, coordinates, width, height, and shape are controlled by processing. In the combination pattern of the block we used the

Fig. 10. Pictures actually projected

coordinates acquired by Kinect and the width and height of the block. Thus, we succeeded in projecting images matching the shapes formed by combining blocks (Figs. 10 and 11).

Fig. 11. Photograph with reduced brightness of camera a state in which an image is actually projected

- Results obtained in this study

By using Kinect, it was possible to recognize the shape of the three-dimensional (3D) space and project the image matching the place of the object, and it became possible to expand the range of expressions as projection mapping and experience.

Also, it succeeded in image projection corresponding to shape change of block in real time, realized interactive projection mapping which can be applied to projection mapping.

Although general projection mapping is often performed outdoors, this work is to be utilized in indoor projection mapping in order to map to the block. In the indoor projection mapping, in terms of installation of equipment, it is possible to reduce the cost compared with outdoors, and by permanently installing it, it becomes possible to easily view and listen.

Furthermore, by eliminating indoor lights, many advantages such as projection mapping can be performed even during the night. Since blocks substitute the things around themselves or are inexpensive and easily available, they are considered to be commercially available as decorations for entertainment supplies for general households, clubs houses and event venues.

- Development and research in the future

Since the object of Kinect can only be recognized at short distance, the shape of the block can be taken only up to 2 vertical and 3 horizontal distances, so the flight distance that can be measured should be increased.

When recognizing an object with Kinect, because the value is acquired with infrared rays, some blurring occurs and the projected image is distorted. Thus, the details of the program should be revised. Since the shape pattern of the combination that projects the image and the image to be projected need to be increased.

Currently only cubes are recognized, so recognizing other shapes such as spheres and cylinders is needed and also to make them correspond so that images can be projected. During projection, the visual aspect shall be developed so that we can project the change of animation when changing the combination, comfortably.

Projection mapping to blocks is an interactive work, and when many users would actually experience it, we would obtain feedback from users' reactions and impressions to further develop them as contents.

We have conducted research with the block parallel to the projector, however, depending on the location there are places where the projector cannot be placed in parallel and we have to resolve it. In addition, currently the block recognition method is correctly recognized only when the projector is arranged parallel to the block. Thus, it is necessary to be correctly recognized according to the position where the projector is placed.

Since it can also detect motion of an object, a function can be added to synchronize sounds when placing an object or moving it. Currently, setting of object recognition is done manually according to the environment to be executed. In the future, we will automate this work and plan to improve so that it can easily be set up by anybody.

References

1. Murayama, S., Saito, J., Suzuki, M.: Visualization of pleasant acoustic spectrum (2016)
2. Masamichi, S., Kotaro, W., Naoya, M.: Interactive mapping simultaneous detection of sound and position (2016)
3. Hashimoto's: Three-dimensional features for object recognition and its surroundings. http://isl.sist.chukyo-u.ac.jp/Archives/Nagoya-CV-PRML-2015March-Hashimoto.pdf. Accessed 17 Nov 2017
4. Ittousai: Xbox One's new Kinect evolved greatly, facial expressions and heart rate are also recognized, six people simultaneously captured the whole body (2013). http://japanese.engadget.com/2013/05/21/xbox-one-kinect-6/. Accessed 17 Nov 2017
5. Saito, K., Miyagi, S.: An attempt to correct coordinate data acquired from Kinect and volume estimation of objects (2014). https://ci.nii.ac.jp/naid/110009766958. Accessed 20 Nov 2017
6. Muramatsu, M.: Gesture interface based on arm pointing direction estimation using Kinect (2015). http://repo.lib.hosei.ac.jp/bitstream/10114/10577/1/13R4135.pdf. Accessed 17 Nov 2017

Development of Recovery of Eye-Fatigue by VDT Works

Ryotaro Kodera$^{(\boxtimes)}$, Ippei Torii, and Naohiro Ishii

Aichi Institute of Technology, Toyota, Aichi, Japan
x16044xx@aitech.ac.jp

Abstract. Visual Display Terminal(VDT) including PC and smartphone and tablet is developed by the computerization technology and becomes the necessities of the modern life now. The number of people who suffer from eye-fatigue because of work using VDT for a long time increase. Doing the works looking at near such as the VDT work for a long time causes eye-fatigue, and ciliary muscle tense, which causes the drop of the sight function. In this study, I developed the application to promote the exercise that relaxes ciliary muscle and ocular muscles using VR animation. I tested it to inspect usefulness of the application. Before VDT work and After VDT work, having watched VR animation after VDT work, I measured the focal distance of eyes and eyesight of the subject. When the focal distance of eyes became long, in this study, it showed that ciliary muscle got tired. I assume this as a "fatigue degree". I developed the application that could recover from eye-fatigue easily by watching 3D animation using VR glasses without using a special apparatus conventionally. This application can expect the asthenopia prevention of the people whose sight function decreased by everyday VDT work.

Keywords: VR · Visual Display Terminal · Smartphone · Eye-fatigue

1 Introduction

Visual Display Terminal(VDT) including PC and smartphone and tablet is developed by the development of the computerization technology and becomes the necessities of the modern life now [1]. Works using VDT becomes it for a long time, and the illness that mind and body called the VDT syndrome become out of condition increases [2]. One of the symptoms is eyes fatigue. Because work to look at the neighborhood such as the VDT works become it for long time, it happens. Ciliary muscle regulating thickness of the crystalline lens is tense to focus it and causes the drop of the sight function by becoming stiff. Also, Ministry of Health, Labour and Welfare recommends moderate stretch in guidelines for labor hygiene management in the VDT works as follows. "before and after of the operation or during operation, it is desirable to perform stretch exercise and relaxation or light exercise" [3]. Therefore we researched and developed the application that it was possible for a stretch exercise to untie the ciliary muscle which was tense for the people whom a sight function decreased by everyday VDT works easily.

© Springer International Publishing AG, part of Springer Nature 2019
I. Czarnowski et al. (Eds.): KES-IDT 2018, SIST 97, pp. 227–236, 2019.
https://doi.org/10.1007/978-3-319-92028-3_24

With VR (Virtual Reality), We can sense the virtual world made with a computer bodily like the reality world. There is the thing called "VR glasses" that stereoscopic vision is enabled by projecting a different image of the parallax to the left and right eyes by watching a picture warped a little output by the display of the smartphone with a lens. VR glass is cheap for commercially available 100–2000 yen. We developed the picture corresponding to this VR glass as application and was intended to restore eyes fatigue easily. We watch a three-dimensional video with VR glass in this study. An effect same as looking at the distance in turn soon was provided and fatigue in the VDT work recovered or tested it. To measure a fatigue degree by the VDT works, before and after VDT works, we measured the focus distance as eyesight and an aim of the fatigue degree and calculated the rate of change. We let the person watch a three-dimensional video with VR glass and measured again. From experimental data, whether the person could revive eyes fatigue by untying ciliary muscle using VR picture was studied.

2 Purpose of the Research

We measured an adjustment function before and after VDT work introduction for 61 VDT workers in the study of the Itabi professor using eyesight and Accommod-opolyrecorder for a newspaper publisher (the device which we put an eye-mark near the perigee and a distant visible point. Accommodopolyrecorder observes an eye-mark nearby and a far-off eye-mark in turn, and can measure adjustment relaxation time at adjustment strain time) (Table 1) [4].

Table 1. Change of the eyesight before and after the VDT works for the newspaper publisher

N = 61			Before introduction	After introduction
5 m	The naked eye eyesight	The right eye	0.47	0.38*
		The left eye	0.43	0.43
	The best eyesight	The right eye	0.86	0.74*
		The left eye	0.85	0.88

After carrying out asthenopia improvement training to 324 primary schoolchildren (158 boys, girl 166) in the study of Hitomi Takahashi and others, there was an effect. The contents of the training have the stick that the Landolt ring reached the tip. We continue staring at the rift of the Landolt ring while bending and straightening an arm. In this way, exercises were performed to untie ciliary muscle looking at neighborhood and the distance in turn [5]. From this, it is revealed that the stretch that continues staring at neighborhood and the far-off object is effective in easing strain of the ciliary muscle. However, individual difference is reflected on length and how to move arms by a person because subject oneself operates a stick. Therefore we intended to get rid of individual difference by using VR animation.

After showing 235 healthy subjects a three-dimensional vision using the tablet which gave a parallax barrier, and measuring the eyesight in before and after audition and the perigee distance in the study of Tatsuya Yamakawa and others, it is thought

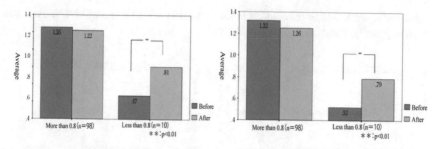

Fig. 1. Change (the right eye, left eye) of the eyesight in before and after training

that there is an effect because eyesight and improvement of the perigee distance were seen [6]. However, there are many brightness falling down to shut out case and screen in itself which do not look like a solid by a viewpoint, and the parallax barrier becoming hard to be seen, problems including the height of the cost of parallax barrier in itself. Therefore the solid showed it based on the above-mentioned fault in our study surely and could find the brightness, and cost decided to show a three-dimensional vision to a subject using cheap VR glasses (Fig. 1).

3   Process of the Study

3.1   Prototype 1

In this study, we developed the application to promote the exercise that relaxes ciliary muscle and ocular muscles by VR animation.

I used Unity3D provided as a free development environment by Unity Technologies company corresponding to 3D expression and the VR output for development. We used VR glass for smartphones to perform stereoscopic vision and installed an animation in a smartphone and tested it (Fig. 2).

Fig. 2. Picture on the smartphone (the left) state that I use (the right)

When an effect same as what watched near and the distance in turn on 3D space by chasing the ball was observed, we expected it when an effect appeared more by giving ocular movement a variation. Therefore, in addition to the picture that a ball worked back and forth, we made four kinds of pictures which added top and bottom, right and left, a turn (spiral), these movement (Fig. 3).

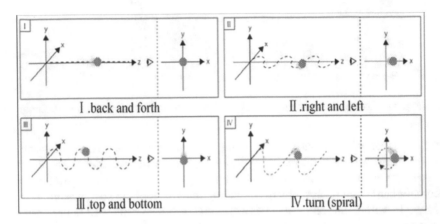

Fig. 3. Orbital sketch of the ball

We tested it to inspect usefulness of the application. After having watched VR animation after VDT works, before and after VDT works, we measured the focus distance and the eyesight of the eyes of the subject three times. VDT works were done for an hour and 30 min, which is a length of the lecture time of the university. Professor Sugita of Shukutoku College advised me about this study, when the focal distance of eyes became long, in this study, it showed that ciliary muscle got tired. We defined this as a "fatigue degree". (We sketch it with "a fatigue degree" as follows).

Professor Sugita also gave advice to me about the method for measurement of the eyesight and performed it based on a standard of the Japanese Industrial Standards (JIS T7309:2002). The following is laboratory procedure about the measurement of the eyesight; by telling a direction of a opend space of Landolt ring pictured in the paper of the square we showed from the 5 m away position in a bright room. The eyesight was in the range of 0.3, 0.4, 0.5, 0.6, 0.7, 0.8, 0.9, 1.0, 1.5, 2.0 (Fig. 4).

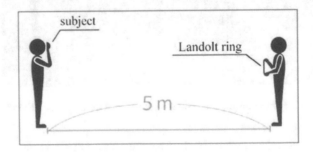

Fig. 4. Figure of the measurement

The following experimental procedure was taken about the measurement of the fatigue degree of eyes (Fig. 5).

- Put the paper with the sentences of the 1 mm-sized characters on a desk.
- The subject brings a face closer to the paper until the sentences become unrecognizable.
- To the distance that can recognize a letter clearly, the subject keeps away a face straight from the paper.
- Measure the distance from paper to the eyes of the subject.

We compare the fatigue with a measurement before the work was performed. Fatigue shows recovered if distance is short, while it shows tired if distance is long.

Fig. 5. Figure of image of the measurement

4 Experimental Results

15 subjects were experimented for the fatigue degree before and after VDT works, in which the eyesight of the subject to see anteroposterior movement with right and left, up and down, and spiral, result in Tables 2 and 3 were experimented. Because of the study was taken over for a few days, a difference produced in eyesight under the influence of weather and temperature by a day even if it is an same person. Therefore, every animation, I took a data of a subject as another person. The eyesight did 2.0 that was the highest with one point of 0.3 that was the lowest with the range of 0.3, 0.4, 0.5, 0.6, 0.7, 0.8, 0.9, 1.0, 1.5, 2.0. From the lowest to the highest, We defined the score as 1 to 10 and checked the increase and decrease of the eyesight with the total value of both eyes. On the Table 2, as degree of fatigue changes, We defined increased degree as recovered, decreased as tired, and same degree as ineffective. In the Table 3, as the sum of the score changes, I defined increased as recovered, decreased as lowered eyesight, and same score as ineffective.

Table 4 or 5 show the number of people that sorted a fatigue degree and the increase and decrease of the each eyesight. The result of the fatigue degree showed a tendency to recover. However, we couldn't find the tendency in the score of eyesight from this study. From this result, we found that there was no relationship between watching the VR animation and the increase and decrease of the eyesight. In addition,

Table 2. Change of the eyesight before and after the VDT works for the newspaper publisher

back and forth

I	II	III	IV
back and forth A	12.5	13	13
back and forth B	11	11	12
back and forth C	18	14	12
back and forth D	9	9.5	8
back and forth E	11	11	10
back and forth F	16	16	13.5
back and forth G	12.5	12	10
back and forth H	14	13	12.5
back and forth I	8.5	8	8
back and forth J	9	9.5	9.5
back and forth K	12	12	13
back and forth L	10	11.5	10.5
back and forth M	8.5	8.5	8
back and forth N	10	11	11.5
back and forth O	12.5	15	13.5

right and left

I	II	III	IV
right and left A	14	14.5	15
right and left B	8	8	7
right and left C	8	9	8
right and left D	14	9	9.5
right and left E	10	10	9
right and left F	11.5	11.5	10
right and left G	13	12.5	14
right and left H	10.5	10.5	9
right and left I	9	10.5	9.5
right and left J	11	11	10
right and left K	11	11.5	9.5
right and left L	8	8	8
right and left M	13.5	14	13.5
right and left N	8.5	8.5	8
right and left O	10	8.5	9.5

top and bottom

I	II	III	IV
top and bottom A	10.5	10	8.5
top and bottom B	9.5	10	11
top and bottom C	14.5	13.5	13
top and bottom D	9.5	10.5	9
top and bottom E	8.5	8.5	7
top and bottom F	8	8.5	8
top and bottom G	9.5	8	8
top and bottom H	10.5	10	10.5
top and bottom I	9.5	10	9.5
top and bottom J	13.5	13.5	14.5
top and bottom K	9	8.5	8.5
top and bottom L	14	16	15
top and bottom M	8.5	9	8
top and bottom N	13	16	14
top and bottom O	8.5	8.5	8

turn(spiral)

I	II	III	IV
turn A	8	8	7
turn B	13.5	14	15
turn C	11	12	10.5
turn D	9	9.5	8.5
turn E	8.5	10.5	9
turn F	9	9.5	9.5
turn G	16	17	15.5
turn H	11.5	13.5	12.5
turn I	11.5	13	10.5
turn J	13.5	14	13.5
turn K	11.5	9	9
turn L	7.5	9	7.5
turn M	12	11.5	12
turn N	13.5	13	13
turn O	13.5	9.5	9

I . subject

II . before VDT work

III . after VDT work

IV . after having watched VR animation

Table 3. Eyesight measurement result every animation

back and forth

i	ii	iii	iv
back and forth A	18	19	20
back and forth B	15	13	11
back and forth C	18	18	19
back and forth D	17	17	19
back and forth E	19	19	20
back and forth F	17	15	17
back and forth G	16	16	17
back and forth H	20	19	20
back and forth I	15	15	15
back and forth J	19	18	20
back and forth K	20	20	20
back and forth L	18	18	18
back and forth M	18	18	19
back and forth N	19	19	20
back and forth O	19	15	15

right and left

i	ii	iii	iv
right and left A	19	20	20
right and left B	18	18	19
right and left C	19	20	20
right and left D	20	19	19
right and left E	20	20	20
right and left F	18	18	18
right and left G	19	20	20
right and left H	18	19	19
right and left I	17	18	18
right and left J	16	17	17
right and left K	18	16	16
right and left L	15	16	15
right and left M	16	12	16
right and left N	20	19	19
right and left O	17	16	17

top and bottom

i	ii	iii	iv
top and bottom A	20	20	20
top and bottom B	16	18	17
top and bottom C	18	20	19
top and bottom D	11	11	13
top and bottom E	14	16	13
top and bottom F	20	19	19
top and bottom G	13	16	14
top and bottom H	18	18	17
top and bottom I	14	16	16
top and bottom J	20	20	20
top and bottom K	13	15	12
top and bottom L	20	20	20
top and bottom M	20	20	20
top and bottom N	13	11	15
top and bottom O	18	17	20

turn(spiral)

i	ii	iii	iv
turn A	17	20	18
turn B	18	18	18
turn C	17	16	17
turn D	13	13	15
turn E	20	20	20
turn F	14	9	15
turn G	18	18	17
turn H	12	15	16
turn I	17	17	17
turn J	20	20	20
turn K	16	15	16
turn L	18	18	19
turn M	14	14	16
turn N	19	20	19
turn O	15	16	15

i . the total value of both eyes

ii . before VDT work

iii . after VDT work

iv . after having watched VR animation

the number of subject recovered from eye-fatigue in three movement was bigger than the movement of back and forth by one. However, many differences were not seen in the movement of the three animations except the anteroposterior movement. From this result, adding movement of right and left, up and down to the anteroposterior movement has some effect, but the most essential is the anteroposterior movement.

According to the result from Table 2, how much effect of the VR animation is

Table 4. Division list of the fatigue degree (Number of people)

back and forth

	Fatigue	No change	Recovery
I	6	5	4
II	3	3	9

right and left

	Fatigue	No change	Recovery
I	5	7	3
II	4	1	10

top and bottom

	Fatigue	No change	Recovery
I	7	3	5
II	3	2	10

turn(spiral)

	Fatigue	No change	Recovery
I	10	1	4
II	2	3	10

I . after VDT work
II . after having watched VR animation

Table 5. Division list of the eyesight (Number of people)

back and forth

	Declining	No change	Recovery
I	1	10	4
II	10	4	1

right and left

	Declining	No change	Recovery
I	7	3	5
II	3	10	2

top and bottom

	Declining	No change	Recovery
I	6	6	3
II	3	6	6

turn(spiral)

	Declining	No change	Recovery
I	4	8	3
II	7	4	4

I . after VDT work
II . after having watched VR animation

shown for the recovery from eye-fatigue by combining the movement of four animations as one. Because of the difference between the initial value of the fatigue degree of subjects, we defined a fatigue degree before the VDT work as 100%. After having watched VR animation, we defined a fatigue degree after the VDT work as 100%. From these, the rate of change of before and after VDT work and before and after having watched the VR animation were measured.

The rate of change of before and after the work [%],The rate of change of before and after having watched VR animation [%], I show it as follows,

The rate of change of before and after the work [%] = Fatigue degree before the VDT work [cm]/Fatigue degree after the VDT work [cm] × 100,

The rate of change of before and after having watched VR animation [%] = Fatigue degree after the VDT work [cm]/Fatigue degree after having watched VR animation [cm] × 100.

The results are shown in Fig. 6. The area surrounded by red and yellow shows that the application was effective. As for the subject who recovered by VDT work, the following reasons are considered; The person had already got tired before a measurement or the person was endurable enough to do the VDT work of one and a half hours. The rate of change of the fatigue degree after having watched VR animation after the VDT works was sorted and was calculated as shown in Fig. 7. We defined the rate of change of the fatigue degree as "restored" if the rate was less than 100%, "not changed" if the rate was 100%, and "tired" if the rate was more than 100%. There was the subject who did not change after VDT work, but the subject who got tired was almost half with 46%. The possibility that the subject didn't changed is that the one had already got tired, and there was the person who recovered after having watched VR animation. After having watched VR animation, about 2/3 of the subject who have done the VDT work were recovered. In addition, the subject who got recovered after the VDT work showed a tendency of getting tired after having watched VR animation. As the cause that got tired after having watched VR animation, eyes may have been cleared by VDT work. It can be thought that eyes got tired by doing the eye-fatigue inspection after having watched VR animation from this state.

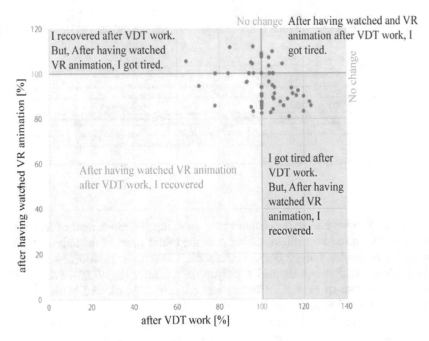

Fig. 6. Rate of change of the fatigue degree to be able to put after having watched VR animation after VDT work

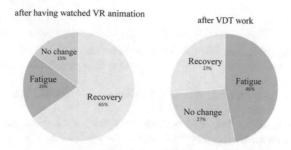

Fig. 7. Ratio of number of people of the rate of change to be able to put after having watched VR animation after VDT work

We developed the application that could restore eyes fatigue easily by watching a three-dimensional vision using VR glass without using a special apparatus conventionally. We can expect the asthenopia prevention of the people whom a sight function decreased by everyday VDT work by using this application. Since the result for the comparison of four kinds of simple movement was a very small amount, We will perform the measurement of complicated movement.

5 Future Development and Research

In this study, there is a possibility that scattering has happened to the fatigue degree of the subject because I measured it after the VDT work of one and a half hours. I want to measure it to obtain a more accurate result when I use application after long-time VDT work. In addition, I want to inspect a more effective usage by measuring a fatigue degree when using it with eyewash, hot eye mask and supplements and also comparing the application with these.

The effect may be differ by the difference of age group because the age of the subjects was 18–21 which can be though to have better resilience than older age group. To solve this problem, I would like to release the application as an application for smartphones and improve it by getting information from users.

I succeeded in the recovery of the adjustment function of eyes by chasing the object with the movement of back and forth with eyes, but room for improvement is anticipated at the point except the movement. Particularly, it is thought that the color has a big influence on human eyes. Various studies about the influence that a color gives a human being such as eye-fatigue and concentration are done. Also the truth that not only an object watching closely but also the background affects to eyes is known. [7] So I change the color of the object, a background and inspect movement and various combinations of colors and will develop a more effective animation in future.

In addition, by using the know-how that I found from this study, ways to recover from eye-fatigue can easily become popular through the amusement such as movies and Smartphone games.

References

1. Ministry of Health, Labour and Welfare: The general condition of the fact-finding result about 2008 innovation and the labor (2008). http://www.mhlw.go.jp/toukei/itiran/roudou/saigai/anzen/08/index.html
2. Sagai, T., Kawadumi, M., Onishi, Y., Ohya, T., Koyama, H.: Relation between Eyeblink and Eyestrain on VDT Work
3. Ministry of Health, Labour and Welfare: Guidelines for labor hygiene management in the VDT work http://www.jaish.gr.jp/horei/hor1-43/hor1-43-9-1-2.html
4. IbI, K.: Accomodation in Technostress Pphthalmopathy
5. Takahashi, H.:: One consideration about the effect of the asthenopia improvement training, About an eyesight improvement effect to look at the neighborhood
6. Yamakawa, T., Tahata, H., Kojima, T., Morita, I., Sugiura, A., Kinoshita, F., Uneme, Ch., Yoshikawa, K., Honda, Y., Mitao, M.: The Effect of Stereoscopic Images of Tablet Devices on the Ease of Eye Fatigue
7. Sato, M.: The influence that a background color gives to problem accomplishment

Author Index

A
An Duong, Thi Binh, 44

B
Bakharia, Aneesha, 1
Bogner, Justus, 109
Brehm, Robert, 150
Bruce-Boye, Cecil, 150

F
Flaegel, Gordon, 150

I
Iijima, Shinya, 54
Inoue, Masahiro, 11
Ishii, Naohiro, 187, 197, 207, 217, 227

J
Jugel, Dierk, 109

K
Keller, Barbara, 120
Kinoshita, Eizo, 166
Kodera, Ryotaro, 227

L
Labella, Álvaro, 76
Lawanont, Worawat, 11

M
Mani, Neel, 128, 139
Martínez, Luis, 76
Menz, Jendrik, 150
Misaki, Hiroumi, 65

Mizuno, Takafumi, 181
Möhring, Michael, 109, 120
Mongkolnam, Pornchai, 11

N
Namugenyi, Christina, 139
Nimmagadda, Shastri L., 128, 139
Nishizawa, Kazutomo, 158
Niwa, Takahito, 187, 197
Nukoolkit, Chakarida, 11

O
Ogikubo, Mateus Yudi, 207
Ohya, Takao, 166
Okada, Isamu, 173
Okamoto, Motoi, 33
Oyamaguchi, Natsumi, 173

P
Pandey, Gaurav, 22

R
Redder, Mareike, 150
Reiners, Torsten, 128

S
Sandkuhl, Kurt, 109
Sato-Ilic, Mika, 33, 54
Schmidt, Rainer, 109, 120

T
Tajima, Hiroyuki, 173
Takahashi, Masao, 33
Toko, Yukako, 54

Torii, Ippei, 187, 197, 207, 217, 227
Tsuchida, Jun, 44
Tweedale, Jeffrey W., 87

W
Wada, Kazumi, 54
Wang, Shuaiqiang, 22

Y
Yabuuchi, Yoshiyuki, 98
Yadohisa, Hiroshi, 44
Yonemoto, Kazuya, 217

Z
Zimmermann, Alfred, 109, 120

Printed in the United States
By Bookmasters